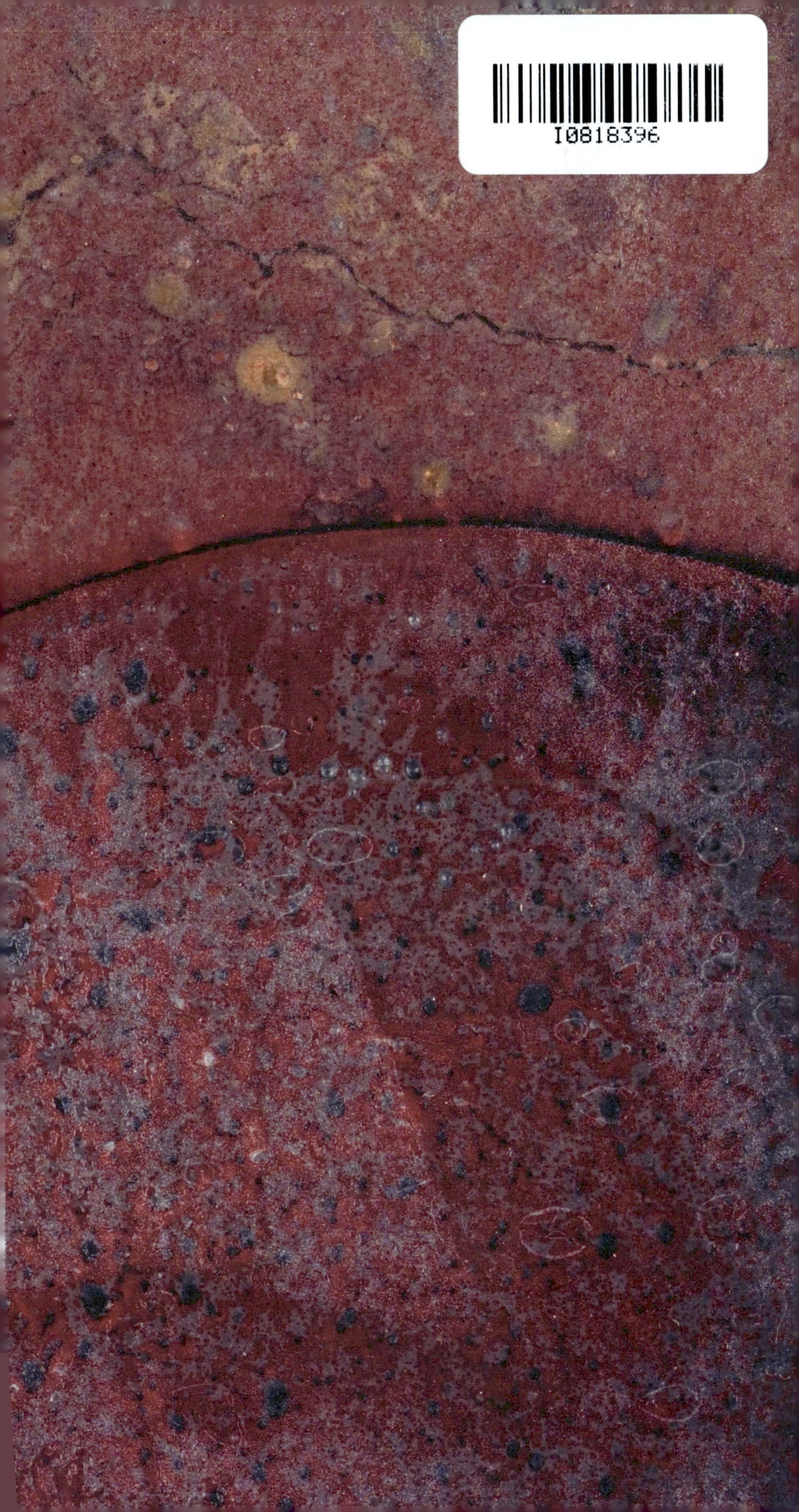
I0818396

Edition Angewandte – Book series of
the University of Applied Arts Vienna
Ed. by Petra Schaper Rinkel, Rector

edition:'ʌngewʌndtə
Universität für angewandte Kunst Wien
University of Applied Arts Vienna

Life is Other

A/Biotic Entanglements in Art and Curating

Edited by Thomas Feuerstein,
Jens Hauser, and Lucie Strecker

DE GRUYTER

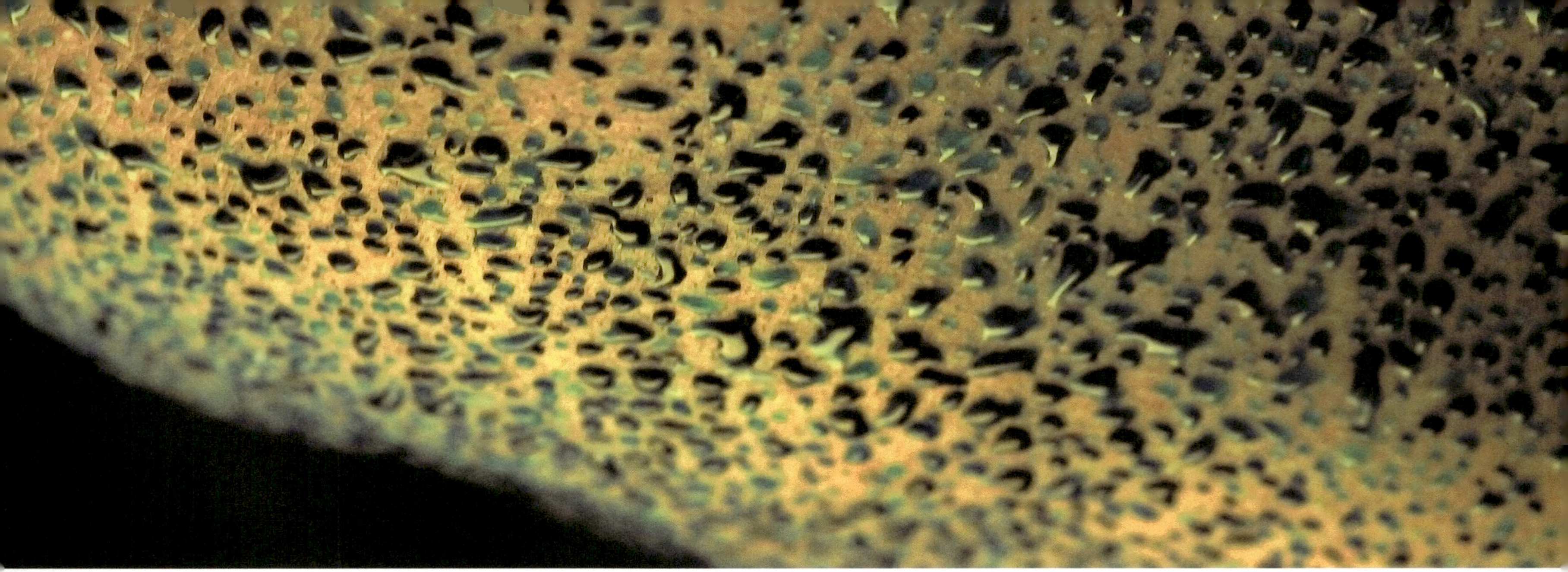

Toward a General "Holobiontics"? Beyond Selfism and Anthropocentrism

Thomas Feuerstein, Jens Hauser, and Lucie Strecker

Fig. 02 Yann Marussich, *Bleu Remix*, 2007, detail from the opening performance of the exhibition *sk-interfaces: Exploding Borders in Art, Technology and Society*, Casino Luxembourg—Forum d'art contemporain, September 25, 2009. © Axel Heise

We humans are inhabited by bacteria, fungi, and viruses, just as we inhabit our own architectures, cities, and environments. Despite all illusions of being in control, we also serve as hosts to the supposedly individual behavioral patterns, ideologies, media, and technologies with which we are intertwined. On the one hand, infinitesimal faulty software security updates cause global tech outages, resulting in grounded flights and media blackouts, but also delays in vital surgeries. On the other hand, movie plots such as Steven Soderbergh's anticipatory *Contagion* (2011), which portrays a worldwide pandemic caused by a deadly virus in the age of global interconnectedness, space-time travel, and technological acceleration, are no longer fiction, but have actually materialized. The social and psychological transformations triggered by the SARS-CoV-2 pandemic, and the increasingly evident consequences of the climate, biodiversity, and energy crises, have made it clear that "liyfe" is first and foremost that of a large variety of agencies that are *other* than human. Simple demarcations no longer stand up to this dynamic. "We" experience "ourselves" as transitory beings drifting between digital and molecular worlds, sensing the twisting of boundaries within us as the emerging possibility of a new language that transcends the symbolic distance between "us" and the world, with an increased potential for multisensory entanglement. How can biological entities, machines, media, architectures, and networks develop symbiotically within the context of contemporary biotechnological possibilities and ecological challenges? And how do the arts transpose this shift, in turn posing new challenges for staging and curating?

Holism is a school of thought that has been advocated since Aristotle in the fourth century BCE and considers the physical, biological, social, and political elements within a system to be interconnected. It is therefore often frowned upon in view of the acceleration that the techno-sciences are experiencing thanks to reductionism and a high degree of specialization. Considering systems in their entirety is often seen to be naïve, obstructive, methodologically shaky, or idealist. However, an increasing number of artists are today focusing on the overlooked and unexpected interconnections between the various components and emergent properties of systems, which cannot be reduced to the sum or differences of their parts. Their positions differ from another trend in the arts at large and the media arts in particular, namely, the negotiation of personal—human—identities as constructed: e.g., queerness is often conceived of as a sociopolitical phenomenon that goes *against* biological determinism, and "nature's queerness" is generally alluded to as an argument for human social role models, but without thinking *with* the biological other. Overcoming the temptation to, e.g., merely update transhistorical traditions of artistic self-portraiture—using an ever-larger variety of media at a time when selfies have become a global phenomenon of individual identity construction within the confines of the human societal structure—artists are raising new questions about binaries such as *self* and *other*, *difference* and *sameness*, *individuality* and *collectivity*, and *carbon* and *silicon-based* worlds. Arthur Rimbaud's famous nineteenth-century declaration "Je est un autre" ("I is another"),[1] which has been interpreted through various lenses (including Lacanian psychoanalysis, as it addresses the fragmented self), suggests that creative impulses stem from psychological splits and external forces beyond the artist's "genius" and cognitive control.[2] Today, however, this "other" actually tends to be physi(ologi)cally explainable and, indeed, to be addressed by the natural sciences in terms of multispecies alliances. What if, for example, we consider that "consciousness—my I-ness—is just the job that the consciousness of my 90 percent bacterial cells and their obligate 10 percent animal multicellular cells require 'me' to do?"[3] In other words, "our planetmates—the others whose space we share and whose shared space, indeed, we 'are.'"[4] Taking this argument even further to the origin of the human sensory nervous system, it could be argued that what

1. Arthur Rimbaud, Lettre du Voyant, to Paul Demeny, May 15, 1871, in *Correspondence inédite (1870–1875) d'Arthur Rimbaud*, ed. Roger Gilbert-Lecomte (Éditions des Cahiers Libres, 1929), 51.

2. See Hauser and Smrekar in their chapter "*HYBRID FAMILY*: An Exchange on the Possibilities of mOtherhood."

3. James MacAllister, in a personal communication with Dorion Sagan, spurred after reading Dorion Sagan, *Cosmic Apprentice: Dispatches from the Edges of Science* (Minneapolis: University of Minnesota Press, 2013).

4. See Sagan in his chapter "Art and Liy͡e, or Carving the Atmosphere."

has evolved over time into today's nerve cells, which allow for "perception, thought, speculation, and memory ... , are the large-scale manifestations of the small-scale community ecology, that is, the fusion of two ancient forms of bacteria."[5]

Emphasizing such symbiotic relationships challenges both human ego concepts and, more broadly, the existing orthodox notions used to define biological individuals, which, in the philosophical sense, then no longer *are* ontologically *in-divisible*, but *become* in that the liѵing modifies its relationship to its environment and also modifies itself.[6] This is at the very heart of the concept of the *holobiont*, popularized from 1991 by evolutionary biologist Lynn Margulis,[7] who described all eukaryotic liѵing beings as the result of mergers between bacteria via reciprocal symbiosis within the biosphere, rather than something shaped by Darwinist notions of competition between organisms and survival through reproductive fitness. While *holobiont* initially referred to the concept of symbiogenesis—e.g., that cell organelles such as mitochondria and chloroplasts were once independent bacteria that have now become part of eukaryotic cells—the concept has been further extended in different contexts. It now applies to virtually all metazoans—human, animal, and plant holobionts with their numerous associated microorganisms, including protists, archaeans, bacteria, fungi, and viruses—and is prone to being metaphorically transferred into cultural and philosophical discourses as well. It is worth noting that Margulis likewise codeveloped the so-called Gaia hypothesis,[8] which posits that the whole Earth functions as a single self-regulating system, where organisms not only coevolve with their environment, but even impact on their abiotic environment, and that the

5. See Clarke in his chapter "Symbiosis and the Holobiont: The Message of Mixotricha paradoxa."

6. According to the concept of co-individuation between a subject and the milieu in Gilbert Simondon, *L'individu et sa genèse physico-biologique (l'individuation à la lumière des notions de forme et d'information)* (Paris: PUF, 1964).

7. Lynn Margulis, "Symbiogenesis and Symbionticism," in *Symbiosis as a Source of Evolutionary Innovation: Speciation and Morphogenesis*, ed. Lynn Margulis and René Fester (Cambridge: MIT Press, 1991), 1–14.

8. Lynn Margulis and James E. Lovelock, "Atmospheric Homeostasis by and for the Biosphere: The Gaia Hypothesis," in *Tellus* 26, nos. 1–2 (1974), 2–10.

environment in turn impacts on the biota. Categorical boundaries between living and nonliving, between *biotic* and *abiotic* realms, are destabilized by metabolic activity when "environment becomes organisms, and organism becomes environment."[9]

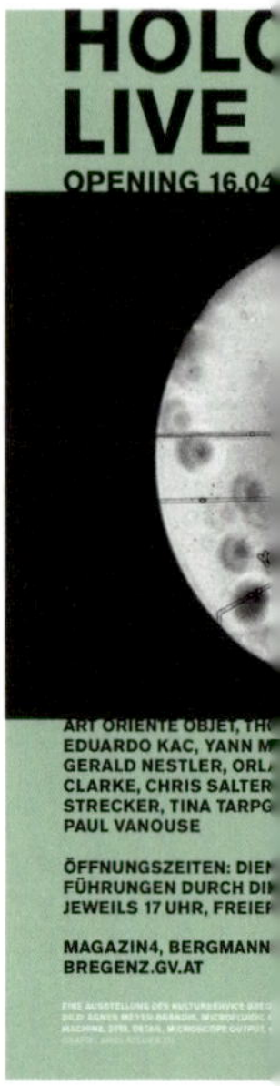

It was for these multiple reasons that we chose *Holobiont* as the title of an exhibition at Magazin 4 in Bregenz that we collectively co-curated with Judith Reichart in 2021 at her invitation. We later presented an extended version of the exhibition at the Angewandte Interdisciplinary Lab (AIL) Vienna in 2022[10], where we shifted the emphasis in *Life is Other* from v to f. In the specific context of coming up with an art exhibition in the midst of the Covid pandemic, the notion of the holobiont corresponded to the tangible planetary relevance of other-than-human agencies for humanity at large, with artists and artworks unable to travel or cross national borders—borders that the virus ignored. Ironically, while the virus, and virality as a concept, had been steady memes and indispensable companions to digital art and cultures in the 1980s and 1990s, in later, rematerialized forms of media art that used various biotechnologies in actual practice, the virus received significantly less attention. For example, it was often reduced to a mere tool for smuggling deoxyribonucleic acid (DNA) sequences into cells—both entities better suited to serving as stable identity proxies than the dynamic virus as a bare code that resists ontologization. As an agent

Fig. 03 Poster for the exhibition at Magazin 4, Bregenz, 2021. © Ahoi Atelier. Image:

9. Myra J. Hird, "Digesting Difference: Metabolism and the Question of Sexual Difference," *Configurations* 20, no. 3 (Fall 2012), 232.

10. Both exhibitions were curated by Thomas Feuerstein, Jens Hauser, Judith Reichart, and Lucie Strecker, with artistic and conceptual contributions from Art Orienté Objet; Irini Athanassakis; David Berry; Guy Ben-Ary; Julia Borovaya; Adam W. Brown; Juan M. Castro and Akihiro Kubota; Tagny Duff; Thomas Feuerstein; Karmen Franinović; Ana María Gómez López; Nigel Helyer; Luis Hernan, Pei-Ying Lin, and Carolina Ramirez-Figueroa; Hideo Iwasaki; Henrik Plenge Jakobsen; Eduardo Kac; Roman Kirschner; Lynn Margulis, Dorion Sagan, Bruce Clarke, and Spherical; Yann Marussich; Agnes Meyer-Brandis; Gerald Nestler; ORLAN; Špela Petrič; Chris Salter; Maja Smrekar; Klaus Spiess, Ulla Rauter, Rotraud Kern, Emanuel Gollob; Lucie Strecker; Tina Tarpgaard; Paul Vanouse; M R Vishnuprasad; Peter Weibel; KT Zakravsky; Adam Zaretsky, and the authors who contributed to the special issue "On Microperformativity," *Performance Research: A Journal of the Performing Arts* 25, no. 3 (2020).

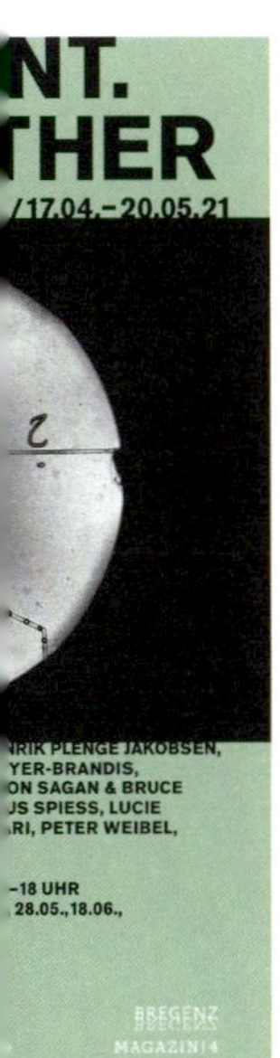

lacking representability and escaping human action, the virus served as challenge and a transmedial trope because, to speak with Jean Baudrillard, it affects us "not as *form*, but as *formula*."[11] While the first staging of *Holobiont* relied entirely on works that could be digitally transmitted "live" rather than physically shipped, the second version, after the hygiene restrictions had been lifted, integrated a number of material artworks and performative installations including various "life forms" and biological agents, all accessible, however, through a 360-degree virtual tour as well.[12] A QR code with a link to the 3D tour can be found at the end of each chapter in this book, allowing readers to access the art works and documentary videos mentioned, displayed in the virtual space. The dynamic QR codes will be updated for future versions of the exhibition. The exhibition examined zones of transition, overlap, and transformation between organisms and their environments through processual artworks that demonstrated liyfe's urge to entangle with others, not only metaphorically but also metabolically, "giving rise to a desire for a narrative of art that, instead of telling stories about the world, makes the world tell stories."[13]

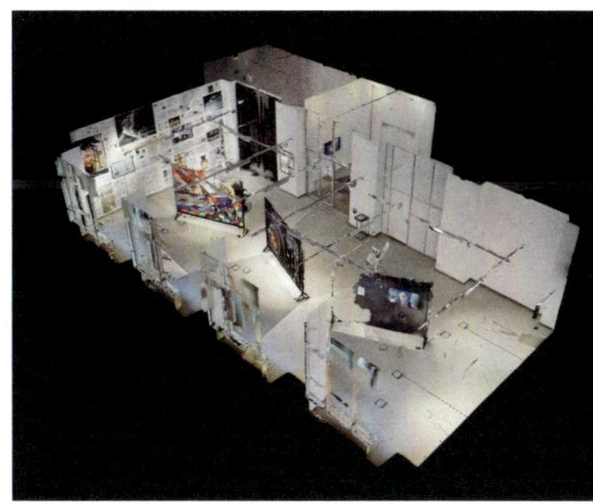

Fig. 04 Matterport Dollhouse View, 2022, *Holobiont: Life is Other*, AIL, Vienna. © Nest Agency

It condensed bodies, environments, texts, media, machines, and biological organisms into multimedia and pictorial spaces, each of which engaged with a narrative about another liyfe and about the liyfes of others. This was where the glyph yf designed for this publication came from. It expresses liyfe's multiplicity as all other liyfe while simultaneously alluding to how the genres of

11. Jean Baudrillard, *Screened Out* (London and New York: Verso, 2002), 1.

12. See QR code at the end of this chapter.

13. See Feuerstein and Holzheid in their chapter "From Symbols to Metabols: Capacity for Synthesis in the Visual Arts."

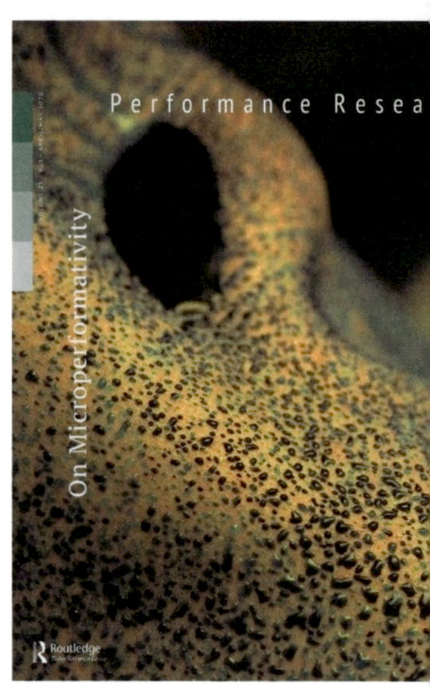

Fig. 05 On Microperformativity, 2019, cover of *Performance Research: A Journal of the*

performance or "live arts," time-based and media arts, have been entangled[14] and expanded to increasingly include not only human bodies but a much broader spectrum of *microperformativity*. This neologism[15] denotes the "current trend in theories of performativity and performative artistic practices to destabilize human scales (both spatial and temporal) as the dominant plane of reference and to emphasize biological and technological microagencies that, beyond the mesoscopic human body, relate the invisibility of the microscopic to the incomprehensibility of the macroscopic."[16] Accordingly, *Holobiont: Life is Other* was framed by a large mural of the complete special issue "On Microperformativity,"[17] encouraging visitors to climb up on stools or ladders, to crawl in unusual positions on the floor, or to take any kind of non-central position in order to get in touch with "holobiontic narratives"—cognitively as well as bodily.

The artists featured in the exhibition made up a large panoply that went beyond the "traditional framework of symbols, metaphors, and allegories, ... to transcend ancestral iconic and linguistic domains."[18] They traced the connections between deforestation and zoonoses in the spread of contagious viruses; interacted bodily with plant growth and chloroplasts in durational performances; taught freedom to stones in a classroom; conducted immunological

14. Chris Salter, *Entangled: Technology and the Transformation of Performance* (Cambridge, MA: MIT Press, 2010).

15. Jens Hauser, "Molekulartheater, Mikroperformativität und Plantamorphisierungen," in *Wahrnehmung, Erfahrung, Experiment, Wissen: Objektivität und Subjektivität in den Künsten und den Wissenschaften*, ed. Susanne Stemmler (Zurich: Diaphanes, 2014), 173–89.

16. Jens Hauser and Lucie Strecker, "Editorial: On Microperformativity," in *Performance Research* 25, no. 3 (2002): 1.

17. Jens Hauser and Lucie Strecker, eds., "On Microperformativity," *Performance Research* 25, no. 3 (2020).

18. See Feuerstein and Holzheid in their chapter "From Symbols to Metabols: Capacity for Synthesis in the Visual Arts."

self-experimentation to literally "become animal"; enacted hybrid trans-species families; choreographed internal physiological flows instead of physical gestures; co-cultured skin cells from humans of different ethnic origins and other animal species; created proto-cells out of waste materials or extraterrestrial organic matter; let crystals grow in real time on a performer's skin; trained yeast cells' memory function; turned breast milk's microbial constituents into economic currency; transformed fecal matter into biopoliticial pharmakons; grew a "rockstar in a Petri dish" out of biological neurons; and transposed the invisible mechanisms of high-frequency trading into choreographed sensory experience ... to name but a few. All the more, the issue of how to curate microperformativity and otherness increasingly became a question of adequate media and forms of mediation. Of course, the audiovisual documentation of performative actions, the display of physical remnants, and digital 3D-animated restitutions are established standards in media art. But would biologists not conceive of media quite differently? Video or tablet computers aside, would they not tend to think of a thermocycler as a kind of molecular photocopying machine, or of growth media for cell cultures, an artificial milieu composed of both biotic and abiotic factors? Concepts of mediality mutate in the light of collaborations between artists and scientists. What would a biological "like" resemble? The *Close Reading* (2021) station proposed by David Berry and Lucie Strecker invited visitors to select and comment on passages from the journal issue "On Microperformativity" by speaking into what looked like a microphone. However, the device did not record speech but the reader's breath and the microbes it contained—a display reminiscent of and referencing Thomas Feuerstein's *Mikrobograph* (2002), a modified Hasselblad camera replacing light exposure with aerial exposure,[19] freely adapted from William H. F. Talbot's "pencil of nature."[20] In *Close Reading*, instead of the text's details, nuances of meaning, or linguistic effects, the microorganisms associated with the speech act took the stage, growing on the culture medium for the duration

19. In German, this results in a pun, from "Be*licht*ungszeit" to "Be*lüft*ungszeit."

20. William H. F. Talbot, *The Pencil of Nature* (London: Longman, 1844/1846).

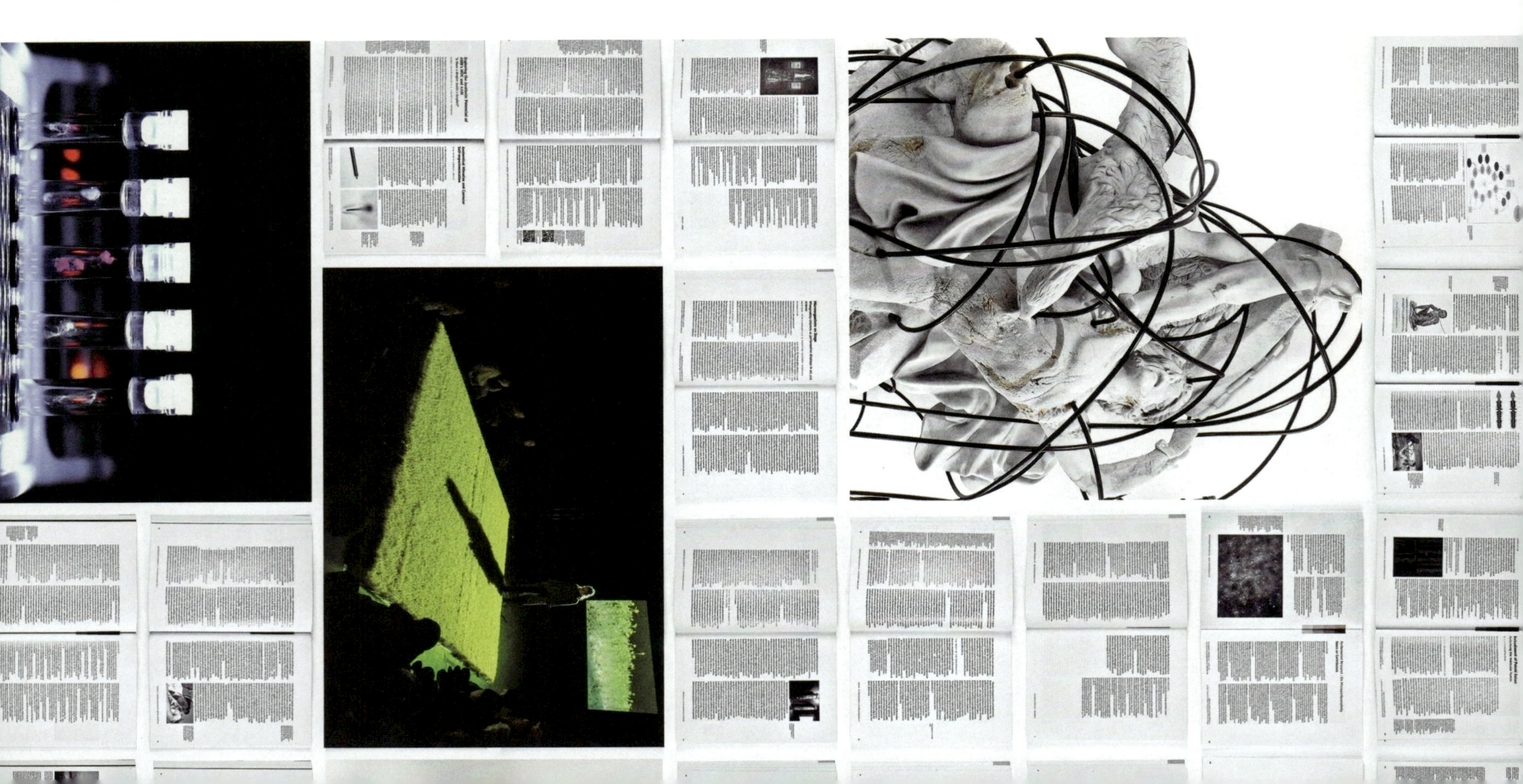

Fig. 06 Jens Hauser and Lucie Strecker, *On Microperformativity*, 2021, newspaper mural, graphic design by Peter Oroszlany. © Lucie Strecker

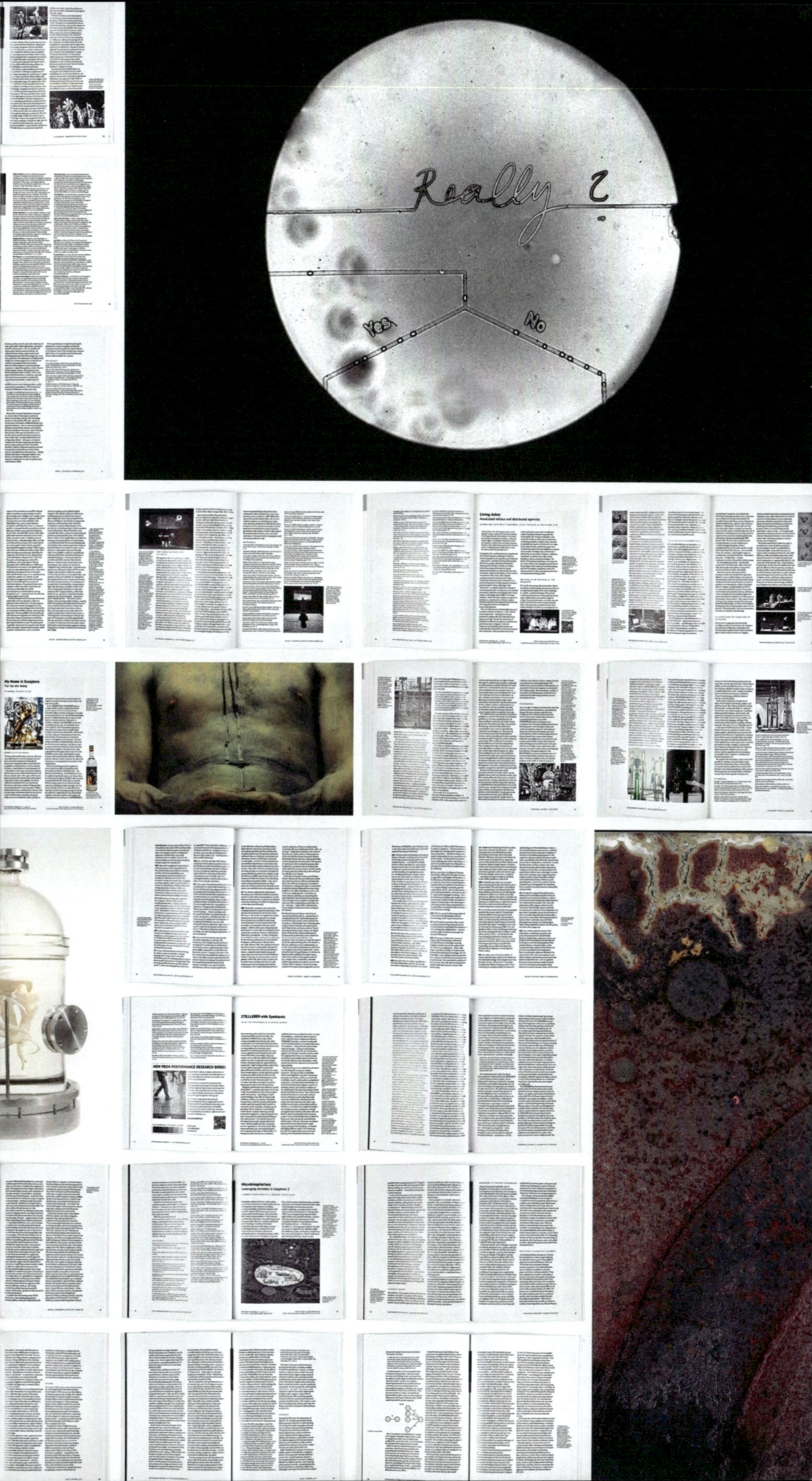
Yes
No
Living Ashes
My Home is Sculpture
STILLLEBEN with Symbionts
NEW FROM PERFORMANCE RESEARCH BOOKS
Microbiospherians

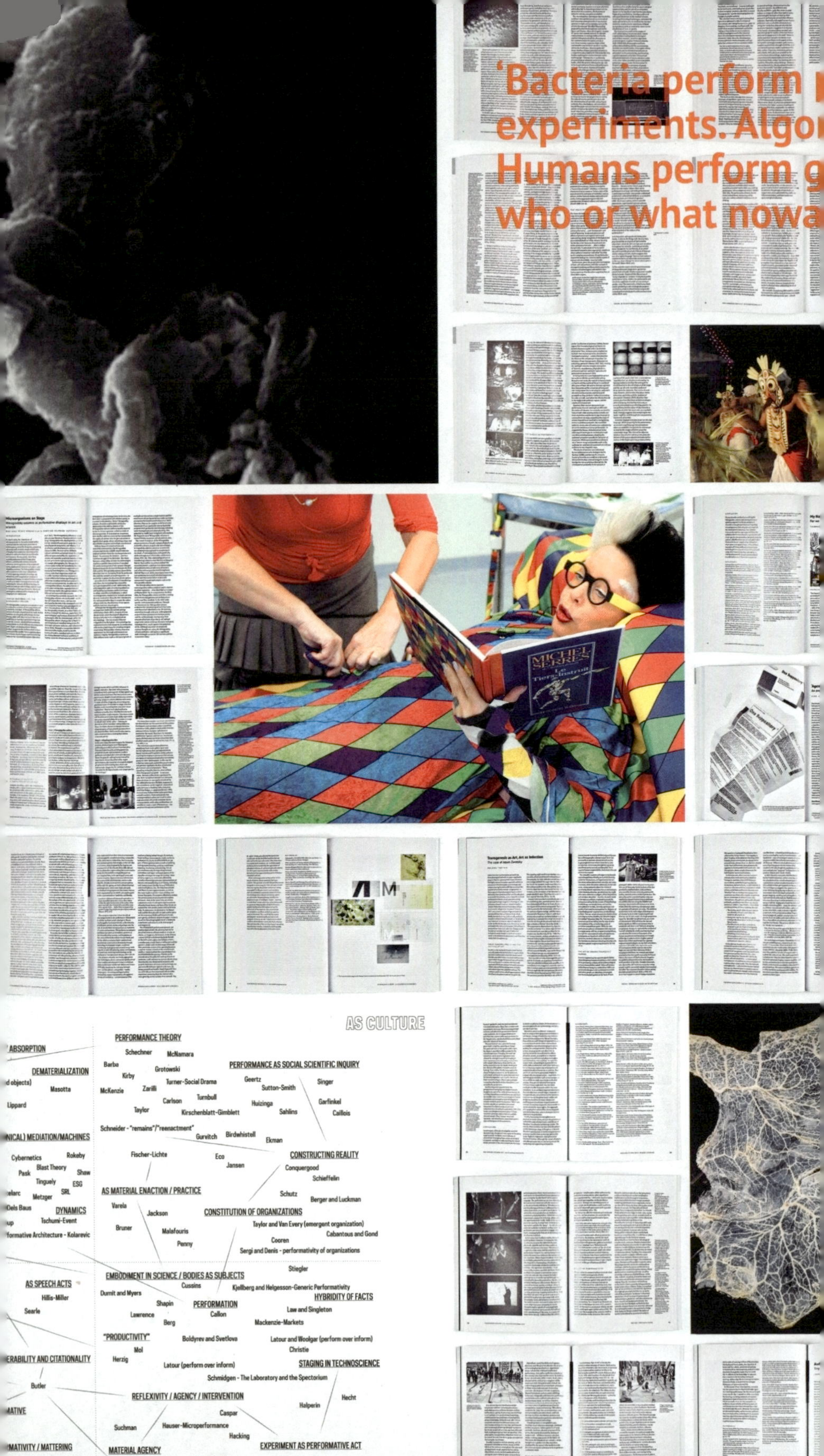

'Bacteria perform
experiments. Algo
Humans perform g
who or what nowa
AS CULTURE
ABSORPTION
DEMATERIALIZATION
d objects)
Masotta
Lippard
NICAL) MEDIATION/MACHINES
Cybernetics
Rokeby
Blast Theory
Pask
Shaw
Tinguely
ESG
SRL
elarc
Metzger
Oels Baus
DYNAMICS
Tschumi-Event
up
formative Architecture - Kolarevic
PERFORMANCE THEORY
Schechner
McNamara
Barba
Grotowski
Kirby
Turner-Social Drama
McKenzie
Zarilli
Carlson
Turnbull
Taylor
Kirschenblatt-Gimblett
Schneider - "remains"/"reenactment"
Gurvitch
Birdwhistell
Ekman
PERFORMANCE AS SOCIAL SCIENTIFIC INQUIRY
Geertz
Sutton-Smith
Singer
Huizinga
Garfinkel
Sahlins
Caillois
Fischer-Lichte
Eco
Jansen
CONSTRUCTING REALITY
Conquergood
Schieffelin
AS MATERIAL ENACTION / PRACTICE
Schutz
Berger and Luckman
Varela
Jackson
CONSTITUTION OF ORGANIZATIONS
Bruner
Malafouris
Taylor and Van Every (emergent organization)
Cabantous and Gond
Penny
Cooren
Sergi and Denis - performativity of organizations
Stiegler
AS SPEECH ACTS
Hillis-Miller
Searle
EMBODIMENT IN SCIENCE / BODIES AS SUBJECTS
Cussins
Kjellberg and Helgesson-Generic Performativity
Dumit and Myers
HYBRIDITY OF FACTS
Shapin
PERFORMATION
Law and Singleton
Lawrence
Callon
Berg
Mackenzie-Markets
"PRODUCTIVITY"
Boldyrev and Svetlova
Latour and Woolgar (perform over inform)
Mol
Christie
ERABILITY AND CITATIONALITY
Herzig
Latour (perform over inform)
STAGING IN TECHNOSCIENCE
Schmidgen - The Laboratory and the Spectorium
Butler
REFLEXIVITY / AGENCY / INTERVENTION
Hecht
Halperin
MATIVE
Caspar
Suchman
Hauser-Microperformance
Hacking
MATIVITY / MATTERING
MATERIAL AGENCY
EXPERIMENT AS PERFORMATIVE ACT
Rheinberger
Barad
Pickering
Knorr-Cetina
PERFORMATIVE SCIENCE(S)

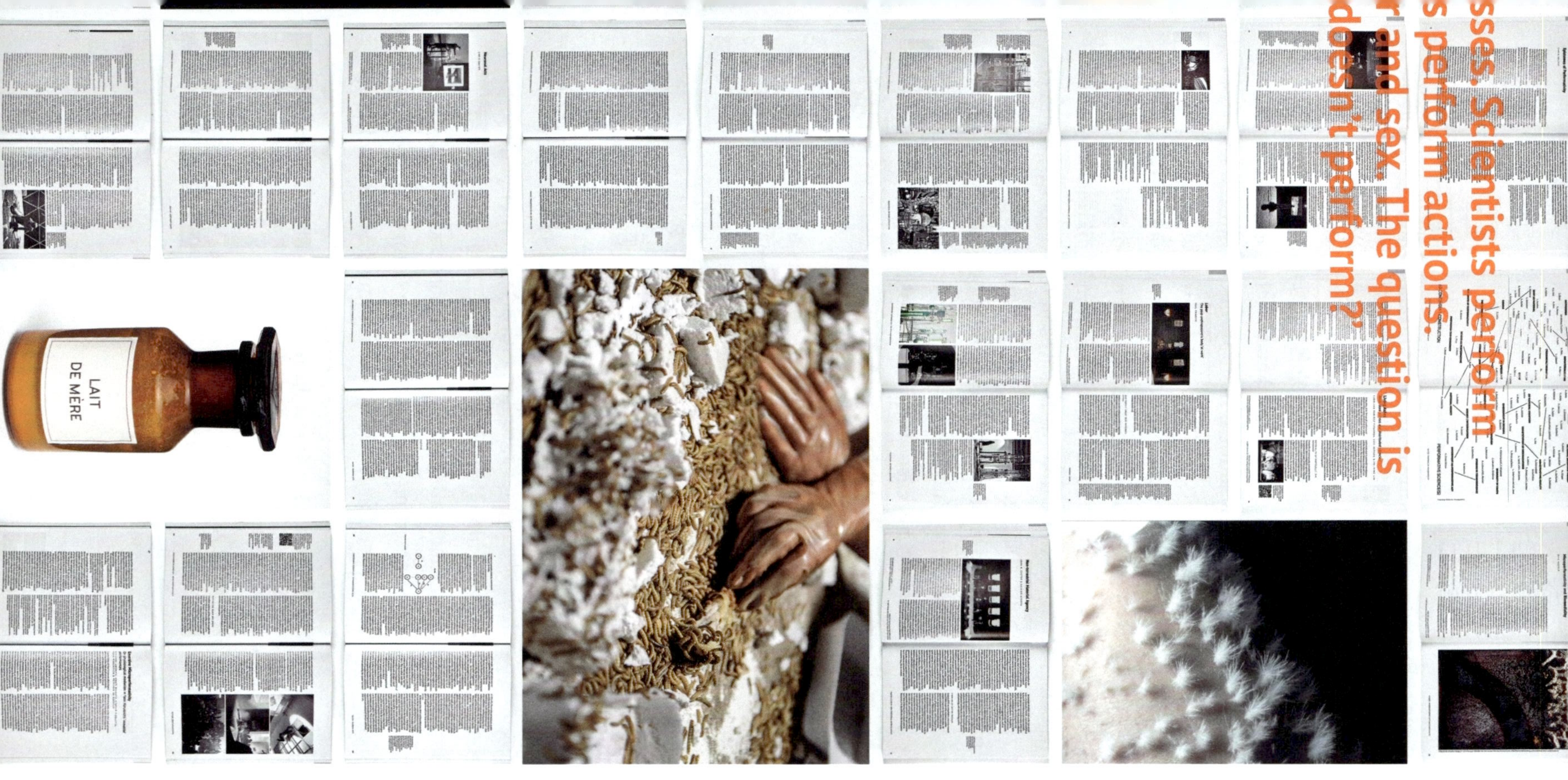

Fig. 07 Jens Hauser and Lucie Strecker, *On Microperformativity*, 2021, newspaper mural, graphic design by Peter Oroszlany. © Lucie Strecker

of the exhibition, prompting reflections on the symbiotic relationship between language and liŷe.

This publication, *Liŷe is Other: A/Biotic Entanglements in Art and Curating*, takes the exhibition *Holobiont: Life is Other* as its starting point, combining theoretical approaches and deepening discussions on selected positions, their specific contexts, and genealogies. It does not propose arguing in favor of fanciful "'biologisms"[21] or risking the misuse of scientific metaphors to cover up the unresolved human social-political challenges of our time, nor does it claim to be an exhaustive compendium of all relevant art projects. Rather, we hope that, with its resolutely eclectic positions, it may contribute to shifting discourses—ranging from "identity," "individuality," and "difference" to "cohabitation" and "entanglement"—and prompt fruitful encounters, such as those described by Bruce Clarke with his case study of the eukaryotic microbe *Mixotricha paradoxa*—a holobiont *par excellence*, without which we might not be able to think or reason today.

The opening chapter by Monika Bakke puts the clear-cut a/biotic distinction to the test by taking a deep-time biological and geological perspective on metabolic forces. She describes how a number of contemporary art practices adopt strategies to stage the complexity of biological and mineral entanglements, linking species through metabolic pathways and networks. Here, "to be entangled is not simply to be intertwined with another ... but to lack an independent, self-contained existence ... . Individuals do not preexist their interactions; rather, individuals emerge through and as part of their entangled intra-relating."[22] Following Karen Barad's argument that matter is not inert or passive, but "substance in its intra-active becoming—not a thing but a doing, a congealing of agency,"[23] with its self-assembling capabilities and metabolic en-

21. E.g., Heinrich Rickert's critique of the systematic transmission of explanatory models from the biological sciences to nonbiological, moral, or social content. See Heinrich Rickert, "Lebenswerte und Kulturwerte," *Logos* 2 (1911/1912).

22. Karen Barad, *Meeting the Universe Halfway: Quantum Physics and the Entanglement of Matter and Meaning* (Durham: Duke University Press, 2007), ix.

23. Karen Barad, "Posthumanist Performativity: Toward an Understanding of How Matter Comes to Matter," *Signs: Journal of Women in Culture and Society* 28, no. 3 (2003), 822.

tanglements, Bakke presents cases in which mineral species have coevolved through liy͡e, e.g., mineralization and coral reefs. In all of her case studies, shared metabolisms are the entangling forces in both biological and mineral evolution.

A similar argument is pursued in the conversation between Thomas Feuerstein and Anett Holzheid, "From Symbols to Metabols." This chapter revolves around the performative installation *METABOLICA,* where bacteria provide the material for sculptures, performing diverse carbon cycles. The Greek term *metabolē* and the German *Stoff-Wechsel* are taken literally—as a profound material transformation in biological and economic as well as artistic terms. "Materials, chemical reactions, living organisms, algorithms, and data streams are becoming actants that are expanding authorship and trying out new aesthetics" for transdisciplinary narratives, along the lines of, but also going beyond, human myths. "Nonhuman entities are becoming the protagonists in new narrative styles that are engaging with the design of technical, social, and ecological spheres"—here, in particular, in order to narrate, metonymically and materially, the shift from petrochemistry to biochemistry.

As a renowned connoisseur of Lynn Margulis' work on symbiosis and the holobiont, Bruce Clarke focuses on one of the main "poster protists for symbiogenesis" in order to speculate on the human ability to speculate,[24] and the biological bases for cognitive capacities in general to depend on the progressive evolution of a rare consortium called *Mixotricha paradoxa,* in which a protistan eukaryote is mixed up with radically different strains of bacteria. Unfolding like a *mise-en-abyme,* Clarke describes how *Mixotricha's* presumed role is potentially at the origin of our sensory nervous systems today, making us "see that, in the biosphere altogether, complex liy͡e persists through symbioses all the way up to Gaia as a planetary system and back down to 'microbial dark matter.'"

In a similar vein, Eduardo Kac aims to translate nonhuman agencies in the biosphere into the aesthetic and interactive,

24. Lynn Margulis, "Speculation on Speculation," in *Speculations: The Reality Club,* ed. John Brockman (New York: Prentice Hall, 1990), 157–67; reprinted in Lynn Margulis and Dorion Sagan, *Dazzle Gradually: Reflections on the Nature of Nature* (White River Junction, VT: Chelsea Green, 2007), 48.

three-dimensional displays of Winogradsky columns, expressing different colors and shapes prearranged by the artist but then left over to gradual metabolic exchange. On the one hand, in his *Specimen of Secrecy about Marvelous Discoveries* (2006), bacteria act as partners or even cocreators, which is well expressed in Kac’s previous formula of an “art that looks you in the eye”;[25] on the other, he raises the question of *for whom* performative modes of nonhuman agency are actually staged, and whether human audiences might not ultimately be the only targets. Human caretakers are required to provide light and nutrient-rich media *to* the perceiving microbial communities. Similarly, the constantly evolving li*f*ing motifs produced *by* them become, in turn, a microperformative aesthetic process perceivable by humans.

Human work, then, can be fully replaced by microorganisms as collaborative, symbiotic agents, employed in Paul Vanouse’s installation *Labor* to shape the olfactory profiles of “our” human identities. In this post-anthropocentric, scientific theatre, it is not individual human identities—cultural, social, or racial—that are responsible for creating the scent of human sweat; rather, skin microbes “fulfil a highly cultural role in non-verbal cueing and signification.” As in the artist’s previous career, where he subverted molecular biological apparatuses—usually employed to analyze differences—to synthesize sameness, in *Labor,* he fully delegates the construction of human individuality to the metabolism of “our” microbiota.

Could there be anything intrinsically more human than speech, language, and memory? “How oral microbes and speech intertwine” is at the heart of Klaus Spiess’ attempt to extend the linguistic concept of “speech acts” to encompass their ultimate materialization. By shining the spotlight on oral microbes as the cohosts of human speech, they push back against the traditional notion of language as an exclusively human domain. Phonemes that are a priori semantically insignificant are used to “train” oral microbes. As the smallest entities of a spoken word, the phonemes,

25. Eduardo Kac, “Art that Looks you in the Eye: Hybrids, Clones, Mutants, Synthetics, and Transgenics,” in *Signs of Life: Bio Art and Beyond,* ed. Eduardo Kac (Cambridge, MA: MIT Press, 2007), 18.

in turn, leave traces of memory on single-celled oral microbes sourced from speakers' mouths, ultimately resulting in "phonetic probiotics."

Hormones are at the center of a conversation between curator Jens Hauser and artist Maja Smrekar about a durational performance of trans-species mOtherhood, *HYBRID FAMILY*. Here, the interplay between two hormones (prolactin and oxytocin), produced in the artist's body through breast-pumping, in combination with a galactagogue diet to promote lactation, enabled her to breastfeed and raise a puppy: "I became Other as mOther." In this project, the coexistence and co-domestication of humans and dogs was addressed via genetic and hormonal coevolution in the light of possible scenarios of cross-species hybridization. At the same time, language-based symbolic categories—such as giving names to nonhuman animals—inscribed the project within the wider history of performance art, including bodily self-experimentation, biosemiotics, and the philosophy of deconstruction.

Going further up in scale, Gerald Nestler reports on the mixed-reality art projects that he conceived together with Sylvia Eckermann, *Planetary Skins* and *Like a Ray in Search of its Mirror*. Likewise inspired by Lynn Margulis's theories, these projects were a "speculative exploration of planetary interconnectedness together with emerging cosmologies and cosmotechnics, and the application of postdisciplinary methodology, bringing together storytelling, conversation, performance, VR and other digital media, video, sound, installations, and other artistic, discursive, and technological tools." The aim was to "encourage more-than-human awareness and ecological solidarity," as well as planetary interconnectedness. At the same time, the artists lament the omnipresence of ambiguous buzzwords and, as Patricia Reed, for instance, puts it, the "fallacy of importing terms like *entanglement* and assuming they have some kind of moral predisposition upon our artificially constructed worlds." Instead, they identify temporal and spatial orientation as the main problem: "Our bodies are volumes of differing magnitudes made up of innumerable entangled scales within a collective liying being. It depends on the resolution inhabited whether the human body is perceived as a soft and solid whole, as humans do, or, as a vast region to traverse and nest in, as microbes might do."

Finally, Dorion Sagan, who is continuing the work he began with his mother, Lynn Margulis, proposes a radically holistic perspective and "project of de-anthropocentrization," observing that "from a broad, cosmic, evolutionary perspective, we are the epiphenomena of multi-billion-year-old and perhaps cosmically distributed, highly durable, and intensely social microorganisms." In addition, some outstanding artists "may have possessed a genius that was partly precipitated by spirochetes; the artists were, in a sense, more than human." Sagan's provocation calls on us to adopt "the broadest view of art," which "includes all liyfe as well as the human artist with or without delusions of personal grandeur." In the context of performance art and microperformativity in particular, where the importance of art objects has been diminished, his conclusion that "the art of liyfe is never static, but exemplifies ... a continuous, three-dimensional flow whose open frames are bodies themselves" resonates with the ambition of this book—"as processes, the artworks of liyfe, unlike those of artists, are never finished." ∞

Digital Content 01 Exhibition walk through *Holobiont: Life is Other*, 2022.

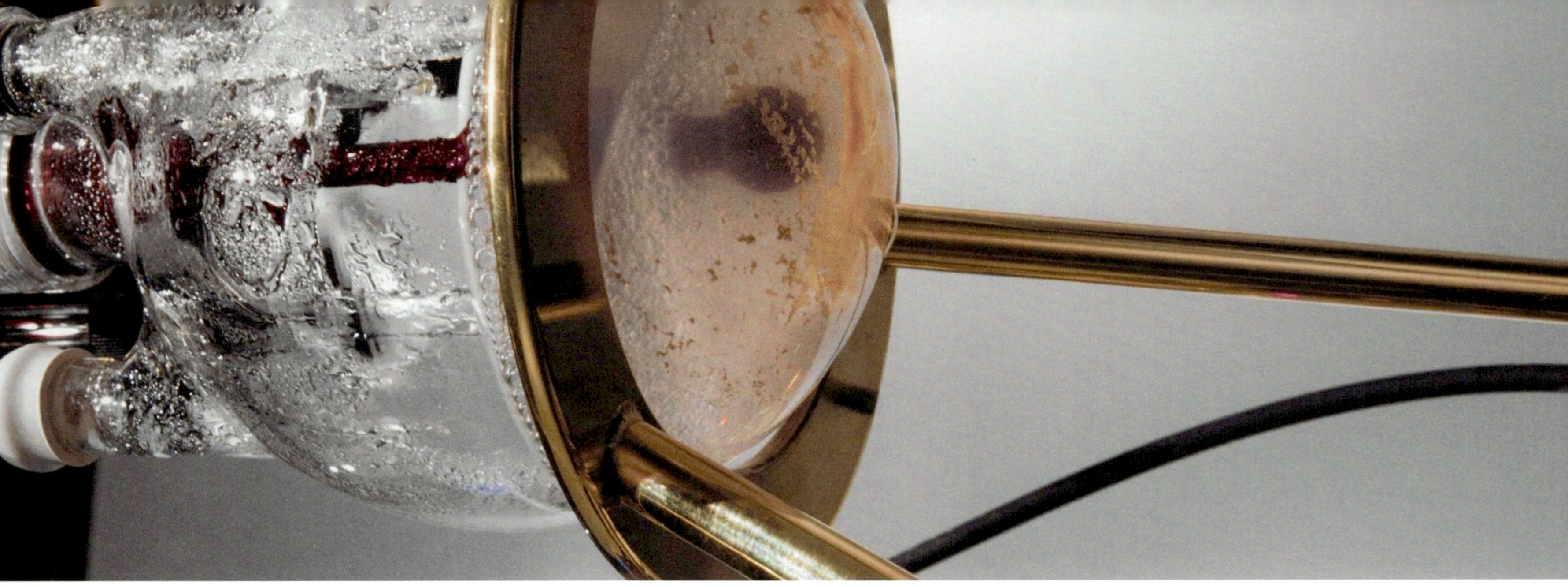

Art and Metabolic Force in Deep-Time Environments

Monika Bakke

Fig. 08 Adam W. Brown, *The Great Work of the Metal Lover*, 2012, gold-plated bioreactor, reducing column, and gassing manifold. © Adam W. Brown

"Metabolism circulates and generates novelty
in all strata of matter/culture."[1]
Myra J. Hird

Contemporary art practices, by taking into consideration not only a biological but also a geological perspective on the living environment and other-than-biological material processes, offer an inspiring means for understanding and communicating the complexity of the biological and mineral entanglements linking species through metabolic pathways and networks. In order to engage with these artworks, it is necessary to adopt Myra J. Hird's rather broad understanding of metabolism as "an evolutionary force; that is, metabolizing as a force of evolution and an inorganic activity through which cells, organisms, life, and nonlife maintain, develop, and respond."[2] This inorganic activity includes processes of self-assembly as well as dynamic forces such as freezing/thawing and concentration/dilution, which enable minerals to diversify and life to emerge, and introduce a whole range of new metabolic possibilities. Geological time, however, is relevant not only when considering key scientific questions concerning the planetary past, such as the chemical origins of life, but also in respect of the planet's future. The latter will be profoundly shaped by the technological challenges being posed by the current environmental crisis. As transformative forces creating disequilibrium and new evolutionary niches, metabolic networks are the focus in nanoscience, synthetic biology, climate science, geoengineering, and many other fields.

Developing an awareness of the overwhelming impact of nonhuman agency on the subjectivity and identity of humans as individuals, as a species, and as carbon-based life-forms requires viewing time from a geological rather than just a biological perspective. Metabolism, as Hird notes, "jolts us from humanist preoccupation to an inhuman evolution,"[3] which goes beyond even the organic. Taking a deep-time metabolic perspective, however,

1. Myra J. Hird, "Digesting Difference: Metabolism and the Question of Sexual Difference," *Configurations* 20, no. 3 (Fall 2012): 234.

2. Ibid., 216.

3. Ibid., 235.

allows us to think of the future of the human species in terms that diverge from the exclusively catastrophic or alarmist. By participating in metabolic networks that include photosynthetic liyfe and mineral species, our bodies reach toward the Sun and the Moon, and, from a deep-time perspective, relate to the cosmic events that resulted in the formation of the chemical elements. A deep-time perspective on metabolic forces reveals an infinity of possibilities for pathways and flows of matter and energy running through both biological and geological realms.

The art practices discussed in this article inquire into the pathways of matter that flow through organic and nonorganic formations undergoing biological and mineral evolutions. At the same time, they contest the boundaries between liyfing and nonliyfing in a mode suggested by Hird, who points out that "through metabolism, environment becomes organisms, and organism becomes environment."[4] The art projects by Michael Burton, Juan M. Castro, Oron Catts and Hideo Iwasaki, Adam W. Brown, and Coral Morphologic analyzed in this chapter from a metabolic perspective investigate the biochemistry of body fluids, protocell membranes, banded iron formations, metal-secreting extremophilic microbes, the calcification of algal and coral reefs, etc. Drawing on both liyfe's mineral origins and the key role it plays in shaping mineral species, artists are turning to technoscience to develop a better understanding of physical, chemical, and biological environments—not just those of the geological past but, above all, those that will come in the future. To address these issues, the artists have developed their artworks in close collaboration with scientists, in some cases creating them in labs. These works operate on various temporal and spatial magnitudes, juxtaposing the human scale with the cosmic scale of celestial bodies, the molecular scale of liyfe's origins, the cellular scale of microbes, and the mega-scale of coral reefs transforming into cities and back again. In both deep time and in every passing moment, the mineral infiltrates all levels of liyfe through metabolic processes.

4. Ibid., 232.

Geology as Genealogy: From Mineral to Biological and Back

Astronomical Bodies (2010) is an art project developed by Michael Burton in collaboration with astrobiologist Terence Kee, who writes that "[c]hemical life would have been the intermediary step between inorganic rock and the very first living biological cell. With the aid of these primitive batteries, chemicals became organized in such a way as to be capable of more complex behavior and would have eventually developed into the living biological structures we see today."[5] Yet, the fundamental building blocks of liýe, such as DNA and ATP, require chemicals like phosphorus, which was only available on Earth in its preorganic stage in a form barely soluble in water and not very reactive. The landing of meteors and interstellar dust, however, is seen as a possible scenario for the arrival of phosphorus minerals much more suitable for transferring the chemical energy necessary to perform metabolic processes.

Drawing on this scientific background, and with a generous amount of humor, the project *Astronomical Bodies* refers to the mineral (chemical) origin of liýe on Earth, viewing it through the physiological chemistry of the human body. But the artist is not interested in tracing the origin of liýe in retrospect, instead seeking to influence the future possibility of catalyzing liýe-forming processes on other planets. Since phosphorus is now not only present in our bodies but also regularly excreted, the artist proposes that it be recovered from his urine with the intention of sending it into space. To visualize this preparatory procedure, Burton has designed an apron with chambers for storing urine and recovering phosphate from the body through a process of crystallization. The phosphate is then intended to be used as a structural component of a man-made meteorite to be sent into space in anticipation of providing a necessary component for the emergence of organic liýe in a new location.

The project *Astronomical Bodies* aims to reconsider the human metabolic relationship with minerals. Burton rearticulates human identity as "P-selves," where the letter "P" conveniently stands for

5. "Power Behind Primordial Soup Discovered," University of Leeds, accessed November 23, 2016, www.leeds.ac.uk/news/article/3386/power_behind_primordial_soup_discovered.

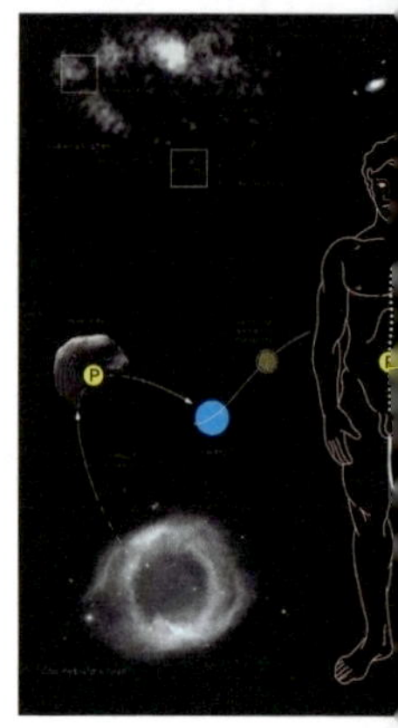

Fig. 09 Michael Burton, *Astronomical Bodies*, 2010. © Michael Burton

both phosphorus and urine. By focusing on a waste substance that all humans excrete regardless of sex—rather than the flow of semen that figuratively depicts life in many cultural traditions—the artist turns something as banal as urination into a life-originating cosmic act. There are two critical aspects to this fantasy of human initiative catalyzing new biochemical origins of life, which takes the form of transposing specific chemicals from a human body into space: First, it stresses the materiality of life. Matter, as Karen Barad points out, "is substance in its intra-active becoming—not a thing but a doing, a congealing of agency,"[6] with its self-assembling capabilities and metabolic entanglements. Second, it posits a radical shift away from the anthropocentric dream of populating other planets with humans accompanied by other species that are only used as tools. This latter scenario has already been well articulated in popular culture, especially in works of science fiction, as well as in the ambitious intentions of nations and organizations to send humans to Mars to establish a colony there. Unlike such space colonization projects, Burton's idea of a new origin focuses on catalyzing a new stream of evolution for carbon-based life, which, in unknown physical and chemical environments, may unfold in a way that is hard to even imagine.

By translating the human body into a flow of matter, the *Astronomical Bodies* project positions the human species in a deep-time perspective that brings up magnitudes far beyond the human scale. Kathrin Yusoff suggests that "the immersion of humanity into geologic time suggests both a remineralisation of the origins of the human and a shift in the human timescale from biological life course to that of epoch and species-life."[7] Burton's immersion into deep time makes a similar argument, but goes much further. He

6. Karen Barad, "Posthumanist Performativity: Toward an Understanding of How Matter Comes to Matter," in *Material Feminisms*, ed. Stacy Alaimo and Susan Hekman (Bloomington: Indiana University Press 2009), 146.

7. Kathryn Yusoff, "Geologic Life: Prehistory, Climate, Futures in the Anthropocene," *Environment and Planning* 31 (2013): 779.

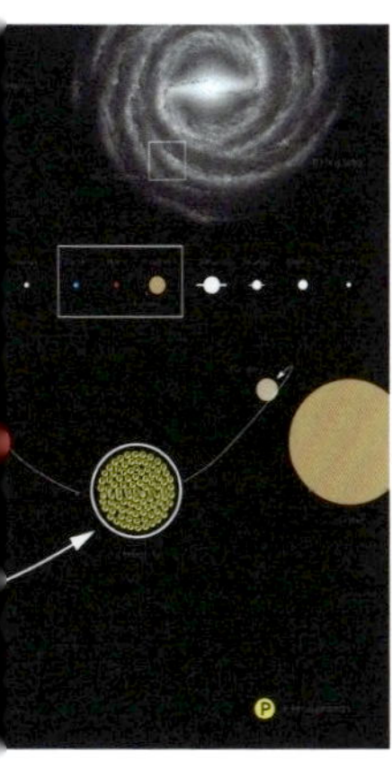

evokes huge intervals of time that not only cause species to disappear but also challenge the demarcation between living and nonliving, organic and inorganic, biological and geological. His interest in species goes beyond biological species, moving toward mineral species and combining biological evolution with mineral evolution.

The idea of a new origin of life envisioned in *Astronomical Bodies* equates human agency with the self-reproduction of carbon-based life. By focusing his work dealing with the origins of life on the biochemistry of the human body rather than the level of genomes, Burton promotes neither individual human genetic identity nor species specificity, but instead inscribes human materiality into a cosmic perspective. The latter is actually a pathway from an abiotic environment to the emergence of life as we know it. In this way, Burton arrives at the mineral edge of life and rearticulates the puzzling question of what life is and how it relates to the geological sphere. As Alaimo pointedly suggests, "Both evolutionary origin stories and contemporary transcorporeal mapping may, perhaps, provoke a recognition that the very stuff of ourselves is always—even across vast scales of times and distance—the stuff of an agential, tangled world."[8] Even prior to the emergence of life on Earth, minerals were diversifying. Although mineral species do not evolve in the same way that biological species do, they come into existence in specific environments shaped by physical and chemical conditions.

However, Burton's artistic gesture toward a new origin induced by the availability of phosphorus falls into the trap of what Carl Sagan called "carbon chauvinism."[9] This means that humans as carbon-based life-forms who have never encountered any life based on alternative biochemistry are inclined to believe that life based on principles other than the one they embody

8. Stacy Alaimo, "States of Suspension: Trans-Corporeality at Sea," *Interdisciplinary Studies in Literature and Environment* 19, no. 3 (Summer 2012): 489.

9. Carl Sagan, *Carl Sagan's Cosmic Connection: An Extraterrestrial Perspective* (Cambridge: Cambridge University Press 1973), 46.

must be in some way inferior. But even liƴe as we know it may have originated more than once in many locations with differing environments; hence, we cannot exclude the possibility that liƴe has emerged in various forms elsewhere. Moreover, there might already exist liƴe-forms with very different types of carbon chemistry, non-carbon-based liƴe-forms, or nonorganic liƴe-forms that go unnoticed due to a lack of instruments and methods that can be used to detect them. But scientists are speculating about alternative liƴe chemistries, positing that arsenic might be a possible alternative to phosphorous. Burton's gesture toward a new origin for carbon liƴe beyond Earth might thus be seen as suggesting a fantasy of the unity of liƴe reflecting some kind of universal harmony. But as Grosz remarks, "the unity of life is not an end, a final harmony or cohesion, but the beginning, the impetus all of life shares with the chemical order from which it differentiates itself, and which it carries within it as its inherited resource."[10]

Perhaps it might, then, be useful to focus on the forces involved rather than on the specific type of matter that liƴe is made from. A working hypothesis of what liƴe is that stresses chemical functionalities rather than specific chemical materials might help humans to think through their carbon biases, including carbon chauvinism. The Program-Metabolism-Container (PMC) model of *minimal life* stresses process and functionality. As Mark A. Bedau argues, "a functional account of life would cover all the chemical systems that share the right chemical functionalities. Physics and chemistry will constrain what kinds of materials can achieve what kinds of functional processes."[11] Liƴe did not suddenly appear in the sophisticated stage of development in which we know it today, but, as David Deamer puts it, "there must have been something simpler, a kind of scaffold-life that was left behind in the evolutionary rubble."[12] "Protocells" are minimal chemical systems, or artificial cells, constructed in laboratories

10. Elizabeth Grosz, *Becoming Undone: Darwinian Reflections on Life, Politics, and Art* (Durham: Duke University Press, 2011), 33.

11. Mark A. Bedau, "The Nature of Life," in *The Cambridge Companion to Life and Death*, ed. Steven Luper (Cambridge: Cambridge University Press, 2014), 23.

12. David Deamer, "Experimental Approaches to Fabricating Artificial Cellular Life," in *Protocells: Bridging Nonliving and Living Matter*, ed. Steen Rasmussen, Mark A. Bedau, Liaohai Chen, et al., (Cambridge: MIT Press, 2009), 19.

from nonliving materials in anticipation of fabricating new forms of liɣ͡e. Although organic liɣ͡e has not yet been developed from inorganic forms, numerous experiments have shown, as Martin M. Hanczyc writes, "the tendency of matter, under the proper conditions, to organize itself into structures possessing similar qualities to living cells."[13]

Minimal Liɣ͡e and Beyond

The discovery of a new origin of liɣ͡e, with its geological and biological complexities, is eagerly anticipated. This search extends to alien environments in space, as envisioned by Burton, as well as to artificial environments within laboratory settings on Earth. Juan M. Castro's art project *Fat Between Two Worlds* picks up exactly where Burton's work leaves off—namely at the biochemical origins of liɣ͡e. Castro is interested in the spontaneous formation of membranes, which allow organisms to distinguish themselves from the rest of the universe and which serve as a key structural element in *minimal liɣ͡e*. Membranes not only contain a cell's molecular machinery but, because they are selectively permeable, also control the movement of substances such as nutrients in and out of the cell, and allow it to capture energy from either light or chemical sources. The metabolism of a liɣ͡ing entity is thus closely related to its membrane, which serves as a container as well as a mediator between the organism and the environment.

The installation *Fat Between 2 Worlds* is based on an *in vitro* experiment that aimed to grow not organisms but nonliving artificial membranes using lipids and self-assembly processes. Castro takes a microscopic view of these processes, which reveal the intricate structures of lipid vesicles self-emerging into elaborate structures in an aquatic environment. The artwork reveals the agency of the inorganic matter and, at the same time, outlines a possible scenario for both the emergence of organic liɣ͡e and the origins of biological evolution. These vesicles built from fatty acids resemble primordial membranes rather than more complex, contemporary membranes built from phospholipids. All organic liɣ͡e that we

13. Martin M. Hanczyc, "The Early History of Protocells: The Search for the Recipe of Life," in *Protocells*, ed. Rasmussen, Bedau, Chen, et al., 14.

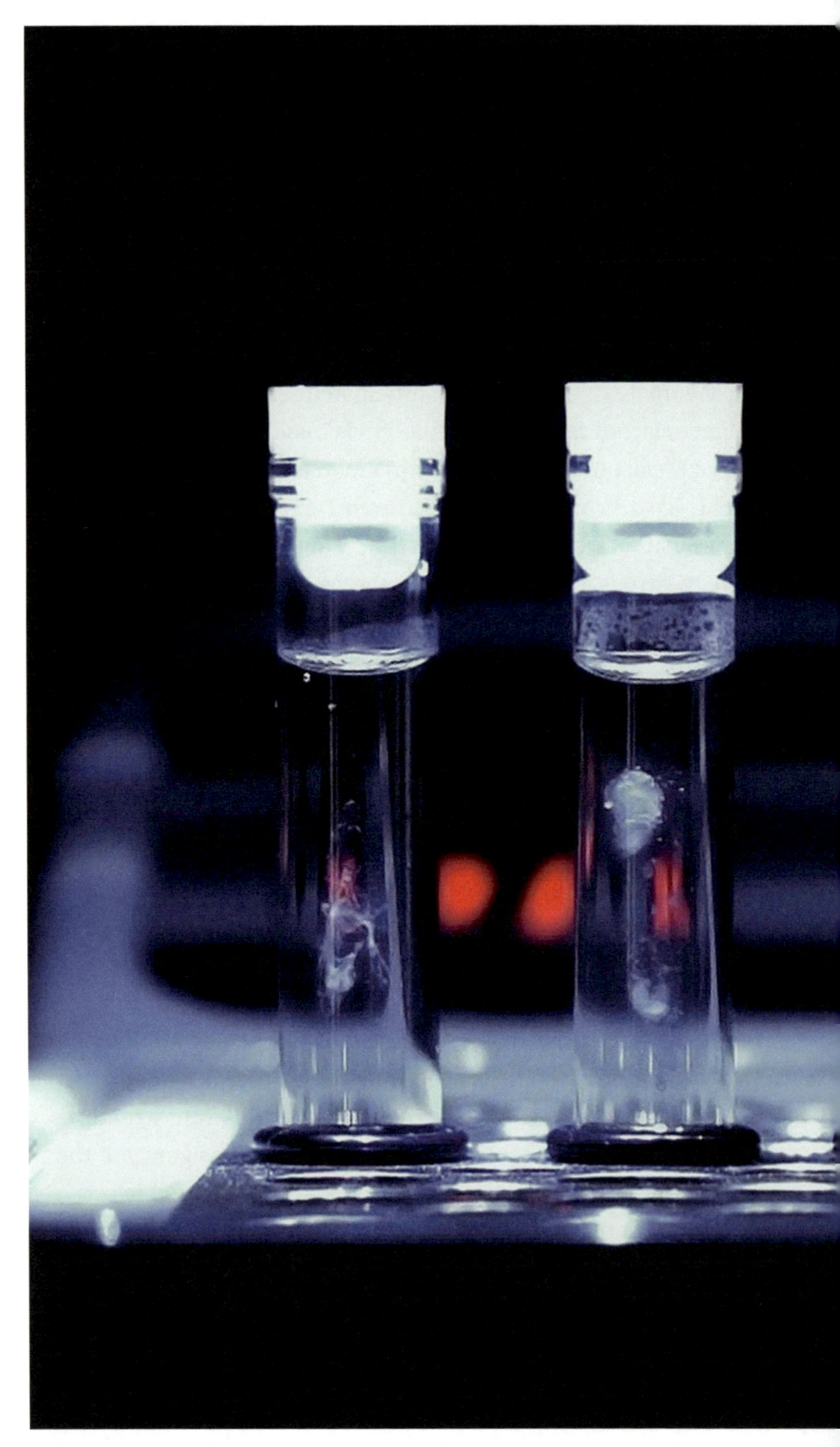

Fig. 10 Juan M. Castro, *Fat Between 2 Worlds*, 2014. © Juan M. Castro

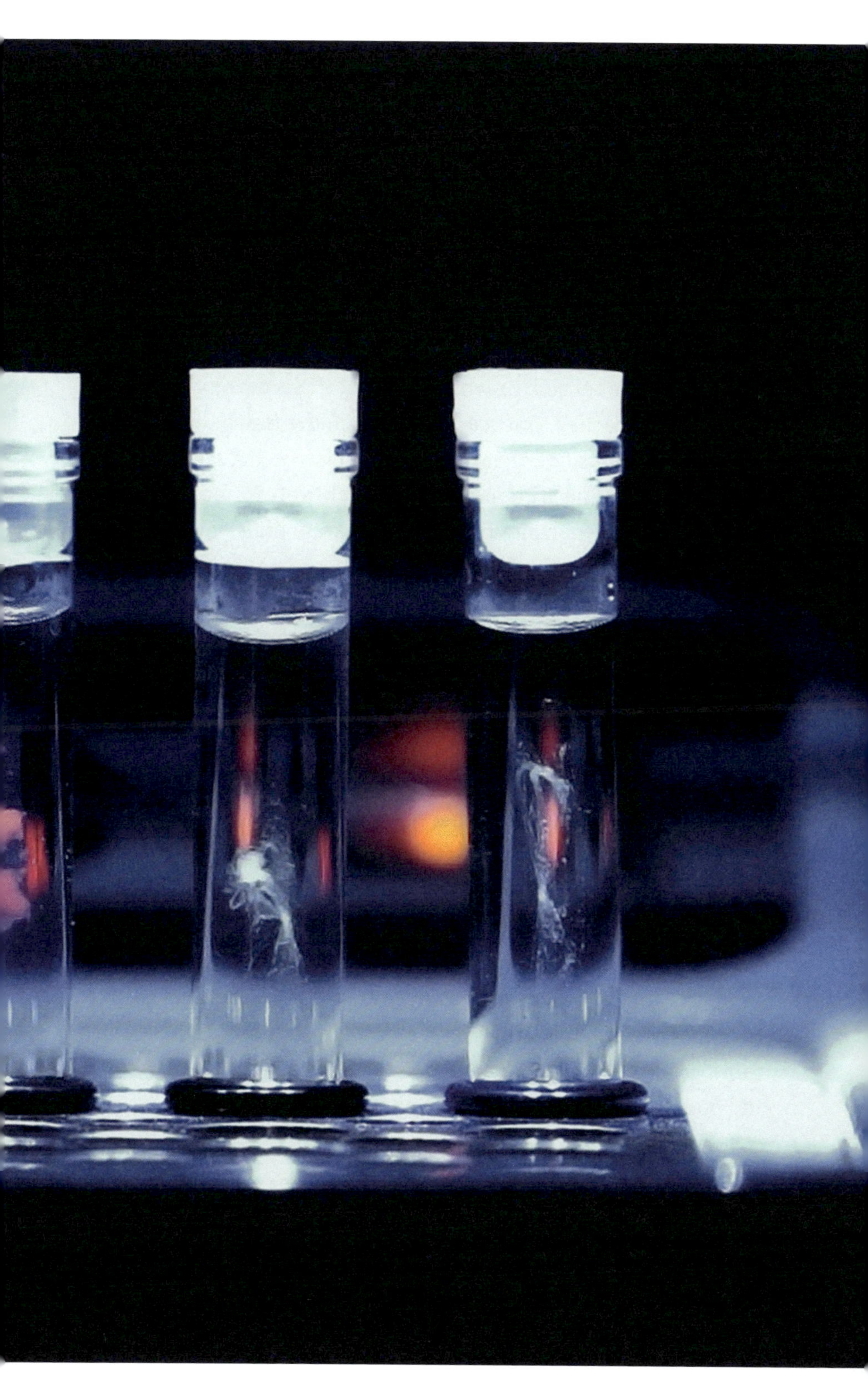

know is cellular; therefore, protocells—the simplest structures located on the very edge of liÿe—are important to scientists because they can tell us something not only about liÿe in the past (in its early stages) but also about designed liÿe in the future.

Finding a coherent pathway from the chemical to the biological is the goal of scientists taking a bottom-up approach to synthetic biology who would like to build liÿing or liÿelike systems from non-living components. Success here would open up endless possibilities for building technological devices that could repair themselves, which could be reused, or which could perhaps develop other desirable properties. One way to approach the task of creating new liÿe is to work *in silico*, that is, in a virtual lab or virtual pot, Castro's method of working with membranes is entirely different. He follows scientists experimenting with the actual materiality of liÿe and liÿelike processes, where the main focus is not on biological information, as in the top-down synthetic biology approach, but on biochemistry and experimenting with self-organizing matter. Castro says that he is "interested in the spontaneous transformation of lipids into soft architectures"[14] and that his project allows the general public to admire aesthetic aspects of these three-dimensional forms with their own dynamics of self-assembly in an art gallery rather than in the relatively inaccessible setting of the lab.

Castro's lipid vesicles are not a product of metabolism in the biological sense, but, as Hird proposes, of a metabolism that should not be limited to organic entities. She claims that "we could equally discuss how hurricanes metabolize wind, currents, air, temperature, and so on. If metabolism is a force, then the demarcation of organic from inorganic and life from nonlife loses traction."[15] Metabolism understood in this way is an evolutionary force in both biological and mineral evolution.

14. Juan M. Castro, "Re-Imagining the Biological Membrane," in *Experiencing the Unconventional: Science in Art*, ed. Theresa Schubert and Andrew Adamatzky (Singapore: World Scientific Publishing Co., 2015), 307.

15. Hird, "Digesting Difference," 235.

Microbial Deep Agency

Life as we know it has a mineral origin, but the origin of many mineral species is closely related to life because they came into existence through biological processes. The coevolution of mineral species with life has been a process of constant interchange. The list of known mineral species is open not only because we have not yet discovered all of them but also because some of them are yet to evolve.

As the simplest life-forms, microorganisms have been dwelling on the planet for the longest period of time. They have also been the most resistant to extreme environments, making them the most influential living agents in the biological-mineral exchange. They have changed the chemistry of Earth's environments on a massive scale, for example, by forming metal deposits like the banded iron formations in Western Australia. Hence, microbial metabolic activity, which has resulted in sedimentary structures, various types of secretions, and oxygen production, should be viewed as a global geological force.

Cyanobacteria are a common form of biological life today, but their deep-time significance and that of their ancestors cannot be overstated. By trapping minerals from water, cyanobacteria form "algal reefs" or so-called "living rocks." These structures, known as stromatolites, are slowly built over long periods of time by successive layers of microbes, with the oldest among them about 3.5 billion years old. As photosynthetic life-forms, cyanobacteria produce waste in the form of oxygen. Thus, when they rose to prominence roughly 2.5–2.4 billion years ago, they changed the Earth's atmosphere into an oxygen-rich one. This was a toxic environment that caused a massive pollution crisis known as the Oxygen Catastrophe, which led to the mass extinction of the anaerobic bacterial life that had been used to dwelling in a carbon dioxide-rich atmosphere. But this oxygenation

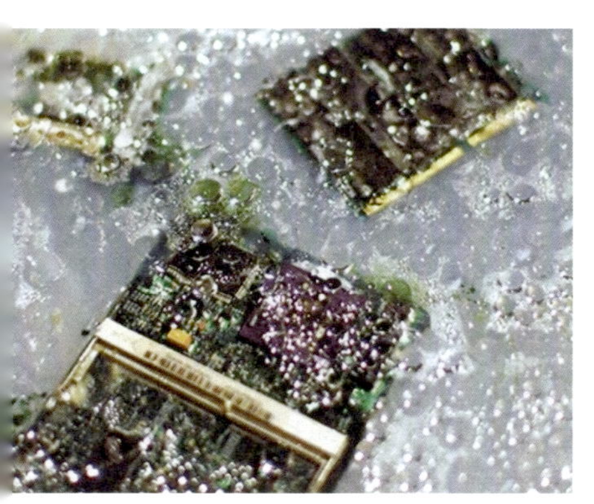

Fig. 11 Oron Catts and Hideo Iwasaki, *Biogenetic Timeslip*, 2010. © Oron Catts and Hideo Iwasaki

event is also viewed as a major evolutionary shift toward the development of eukaryotic cell organisms, including plants and animals. It was also one of the greatest events in mineral evolution.

In their project *Biogenic Timestamp* (2010)—which was realized as a part of a platform called *Synthetic Aesthetics*—artist Oron Catts and biologist and artist Hideo Iwasaki focused on cyanobacteria's metabolic processes, viewed from the perspectives of both the deep past and deep future. The artists, inspired by stromatolites and, in particular, by the metabolic abilities of cyanobacteria "to accumulate, deposit, and precipitate metals and other substances," experimented with "the possibility of human-induced biogenic formation."[16] One outcome of their artistic endeavor was a case study in which computer motherboards were exposed to modified cyanobacteria. The latter were able to "rearrange and deposit some of the components, such as silica, gold, and iron and in this way present a symbolic case in which life 'messes up' the linear logic of digitally driven synthetic biology."[17] But when viewed in a broad sense, metabolic force affects biological, mineral, and other evolutions, including the evolution of technology and art, allowing for the constant transposition of matter. As Hird suggests, metabolism is "an entangling force insofar as it defines a relationship between non-discrete entities, themselves already entangled"[18] in ways not easily detectable. Yet the metabolic pathways of bacteria are perhaps our only hope for dealing with computer waste. Given that discarded computer chips produce layers of minerals, synthetic biology may help to mobilize the already transposed minerals by means of forced evolution, leading to the creation of novel organisms with novel metabolisms.

Microbial precipitation is also the focus of Adam W. Brown, whose art project *The Great Work of the Metal Lover* (2012) refers to both the science of chemistry and the protoscience of alchemy in

16. Oron Catts and Hideo Iwasaki, "The Biogenic Timestamp: Exploring the Rearrangement of Matter through Synthetic Biology and Art," in *Synthetic Aesthetics: Investigating Synthetic Biology's Designs on Nature*, ed. Alexandra D. Ginsberg et al. (Cambridge: MIT Press, 2014), 201.

17. Ibid., 201.

18. Hird, "Digesting Difference," 232.

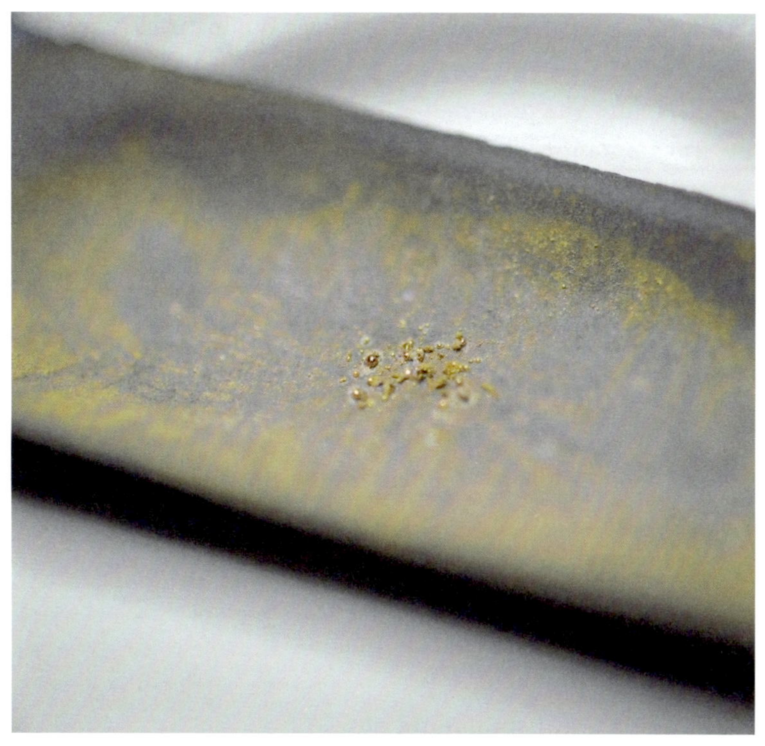

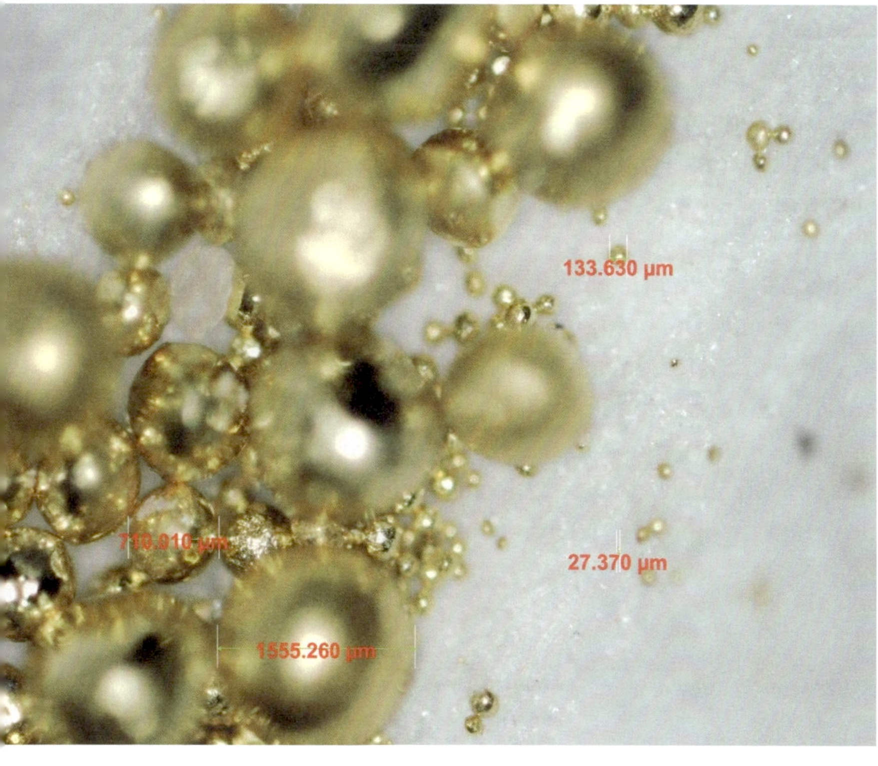

Fig. 12–13 Adam W. Brown, *The Great Work of the Metal Lover*, 2012, biomineralization of gold particles. © Adam W. Brown

its approach to biomediated mineralogy. While the alchemical process of transmutation was supposed to produce gold from ordinary metals, in Brown's project, it is the biological process that performs the transmutation. The key element of the installation is a bioreactor in the form of a vessel containing the extreme ecosystem of an engineered atmosphere and "highly specialized metallotolerant extremophilic bacteria."[19] These were not engineered but found in an anthropogenic toxic environment to which they had adapted. This art installation comprises elegantly presented custom laboratory equipment, providing an environment in which microorganisms can metabolize a highly toxic liquid of gold chloride into pure gold, which is eventually deposited on the bottom of the vessel.

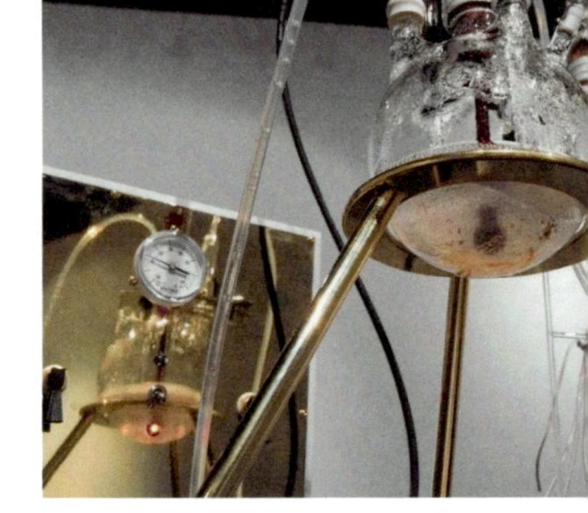

Fig. 14 Adam W. Brown, *The Great Work of the Metal Lover*, 2012, gold-plated

Personal transmutation, purification, and perfection—the alchemist's dream and ambition—were supposed to find their analogy in the physical transmutation of common materials into gold. In many ways, this ambition survives in the contemporary search for purification methods and in the desire to design waste-free metabolisms. The philosopher's stone, also known as the stone of live metal in its contemporary manifestation, turns out to be aliy̆e, and it is not in the hands of alchemists, but in the possession of the mining industry. Microbial mining is nothing new and the microbes involved in the leaching or concentration of metals have been selected from super-adapted survivors, usually found in mining dumps. More recently, however, the mining industry has gained support from synthetic biology, which provides genetically engineered microorganisms designed "to increase their efficiency and allow them to function

19. For more information on the liy̆e-forms involved in this project, see Adam W. Brown, "The Great Work of the Metal Lover," accessed November 20, 2016, www.adamwbrown.net/projects-2/the-great-work-of-the-metal-lover/.

on a larger variety of substrates."[20] biomining is contributing to the human-induced deterritorialization of minerals and participating in "a planetary mine," as Mazan Labban calls it. He believes that biomining reveals a desire to generate "the conditions of a waste-less capitalist mode of production"[21]—unstoppable, boundless, and endless. With microorganisms a new labor force capable of producing new minerals by means of forced evolution, we may wonder about the new ecologies of mineral and organic species still to come in the long run, outside the capitalist mode of production. The novel metabolisms of the designed microbial workforce might be able to generate new mineral species, which will have an as yet unforeseeable imprint on the planet's surface, whether as a niche in which some biological species will flourish or in the form of their toxic waste.

Skeletons All the Way Down

"Skeletons dramatize an ancient waste, whispering a ghostly testimony to the useful internalization of hazardous waste sites," Dorion Sagan points out, explaining that "the calcium phosphate of bone owes its existence to the necessity of eukaryotic cells to keep cytoplasmic calcium concentrations at levels around one in then million."[22] The concentrations of calcium in sea water are much higher, which means that organisms dwelling in the sea must exude calcium to prevent themselves from being poisoned. Mineralization (calcification)—which leads to the formation of stromatolites, coral reefs, shells, exoskeletons, bones, and egg shells, but also to kidney and gall stones—is a trace of the affinity between biological and mineral species. Not surprisingly, then, coral skeletons are highly biocompatible with ours, which is why they are used in human

20. US Congress, Office of Technology Assessment, *Commercial Biotechnology: An International Analysis* (Washington, DC: US Congress, Office of Technology Assessment, OTA-BA-218, January 1984), 226, accessed November 26, 2016, https://ota.fas.org/reports/8407.pdf.

21. Mazan Labban, "Deterritorializing Extraction: Bioaccumulation and the Planetary Mine," *Annals* 104, no. 3 (2014): 573.

22. Dorion Sagan, "Metametazoa: Biology and Multiplicity," in *Incorporations,* ed. Jonathan Crary and Sanford Kwinter (New York: Zone Books, 1992), 368.

bone transplants and as scaffolding in tissue culture. And yet, they are also well suited to building cities that function as human urban exoskeletons. As DeLanda points out, "the human endoskeleton was one of the many products of that ancient mineralization. Yet that is not the only geological infiltration that the human species has undergone. About eight thousand years ago, human populations began mineralizing again when they developed an urban exoskeleton … ."[23] The latter may be built with stone, clay bricks, steel, or concrete, but also with real skeletons.

Marine biologist and video artist Colin Foord and musician Jared McKay form the art group Coral Morphologic, which focuses on the aquatic environment of the city of Miami, often referred to as the "coral city" because, as Foord explains, its "infrastructure is literally comprised of the processed skeletal remains of corals and marine life that once colonized South Florida when it was submerged in eras past." Foord continues: "Almost every building, sidewalk, and highway in Miami contains calcium carbonate-based concrete that is recycled from the remnants of those coral reefs. It is a city where vertebrate and invertebrate life-forms are forever bonded through a calcium carbonate matrix. Skeletons that were once enveloped with fluorescent coral tissue now form the foundation for a neon metropolis to mirror its coral reefs."[24] In response to coral degradation in Florida, Coral Morphologic have produced multimedia and site-specific artworks and also organized coral-related conservation initiatives.

Coral reefs are among the oldest and largest living structures on the planet, although they are composed of tiny polyps. They developed about 251 million years ago following the Permian-Triassic extinction event, which was the most severe mass extinction event the planet has ever experienced. Corals long posed a difficult dilemma to naturalists because they "climbed from the sea's

23. Manuel DeLanda, *A Thousand Years of Nonlinear History* (New York: Zone Books, 1997), 27.

24. Colin Foord, "The Corals are Dreaming Again," accessed November 21, 2016, www.coralmorphologic.com/index.html.

depths to create new landscapes."[25] Their enigmatic status as rocks, plants, or animals was an enduring conundrum, and before French naturalist Jean-André Peyssonnel pronounced them animals in 1753, "coral flowers" were believed to be insects, mollusks worms, or even animalcules similar to microbes. "In the eighteenth-century fascination with the idea of a Great Chain of Being, corals held a special place" as, with their calcified skeletons, they created, as David Dobbs writes, "the huge structures that joined the organic and the inorganic worlds as well as sea and land."[26] In other words, they linked the geological with the biological. Charles Darwin, who encountered reefs on his voyage on the *Beagle,* investigated their landscape-forming ability and presented his findings in his very first book, *The Structure and Distribution of Coral Reefs,* published in 1842. He claimed that the development of fringing, barrier, and atoll reefs was closely related to the geological structures on which they grew and on changing sea levels.

Although coral polyps are animals, they actually live in a symbiotic relationship with tiny organisms called zooxanthellae, which provide them with food from photosynthesis. This vital relationship becomes problematic when corals become stressed and begin ejecting the zooxanthellae from their tissue. This process, known as bleaching, makes coral bodies lose their vivid colors and eventually turn white. It does not kill the coral right away, but this prolonged stress can be fatal. Coral bleaching is now a global phenomenon caused by anthropogenic and non-anthropogenic factors, such as temperature rises, extra-bright sunlight, pollution, changes to seawater salinity, and sedimentation caused by dredging. Climate change is having a strong impact on coral reefs. Among the other types of damage it causes, it disturbs the calcification process. Ocean acidification is intensifying due to the increased absorbtion of carbon dioxide from the atmosphere. The chemical condition of oceanic environments is causing a decrease in the rates at which reef organisms calcify and an increase in the dissolution of the

25. David Dobbs, *Reef Madness: Charles Darwin, Alexander Agassiz, and the Meaning of Coral* (New York and Toronto: Pantheon Books, 2009), 145.

26. Ibid.

reef sediments that form the reef structure.[27] Reef disappearance is leading to a loss of fish habitats but also to costal erosion. It is not possible to predict whether coral metabolism will keep up with changing environmental conditions or how such changes will affect the overall ecosystem.

Some of the videos made by Coral Morphologic are inspired by their own discovery of super-adapted hybrid corals living around Miami's shores. These resilient urban corals of Miami prove that these animals are responsive and adaptable, and are able to cope with heavy pollution produced by the city as well as other negative factors related to the global environmental crisis. With the warning that Miami will actually be one of the first cities in North America to be flooded due to climate change in mind, Coral Morphologic coproduced (with Borscht Corp and Lucas Leyva) a short film called *The Coral Reef Are Dreaming Again* (2014),[28] which makes reference to the possibility of corals reinhabiting Miami in the absence of humans. The video shows the corals that will come after humans as being capable of carrying the physical memory of their past material forms. This includes the architecture of the city and its human inhabitants, whose sexed bodies and gender roles, and therefore their identities, "are fluid and ever-shifting. This is no different from the coral reef, where many species are frequently transsexual or hermaphroditic."[29] In the video, a human body initially undergoes an identity change and, eventually, through submersion, becomes a geological formation, a skeleton, colonized by corals. In this way, the nightlife of the once urban, queer environment continues as queer coral communities. The latter observe the Sun and Moon in order to carry out key metabolic functions like feeding during the day and sexually reproducing at night during a full Moon and at a specific time of year. This is how coral metabolic networks reach

27. Richard B. Aronson, ed., *Geological Approaches to Coral Reef Ecology* (Heidelberg: Springer Science and Business Media, 2007).

28. "The Coral Reef Are Dreaming Again," accessed June 5, 2024, www.vimeo.com/219736461.

29. Colin Foord, "The Corals are Dreaming Again," accessed November 21, 2016, site inactive as at June 8, 2024, www.coralmorphologic.com/index.html.

out in time and space, far beyond both the city of Miami and the human species.

After flooding, human life in the city will certainly discontinue, while robust urban corals may get a chance to colonize Miami's urban exoskeletons as well as human skeletons. Thus, in the underwater environment to come, skeletal matter—the geological—will reassemble itself once again into yet another temporary structure. In anticipation of this process, the artists have been encouraging the building of coral nurseries and promoting coral rejuvenation, relying on the resilience and adaptive abilities of the corals, which are still able to cope with high levels of pollution. Similar efforts have been made in Hawaii, where the scientists from the Hawaii Institute of Marine Biology have gone one step further by actually trying to breed a "super coral" by means of assisted evolution.[30] Corals sampled from a reef are grown in a laboratory setting, where they are exposed to heat, acidity, and light, and the most resilient are selected for further development. The objective is to select and create a coral metabolism capable of withstanding changing environmental conditions.

Coda

Metabolism, as Hird points out, "messes with arboreal evolutionary lineage,"[31] operating as cycles and systems in which mineral and biological species meet constantly, and are necessary companions to each other. Art practices inquiring into metabolic force put forward the concept of metabolism as a process that transforms and decenters, as cosmic recycling. They offer both a glimpse into the geological past and provide inspiration to speculate about the biochemical futures of the Earth and other cosmic environments. But these futures of multispecies alliances have already begun, and they seem to be influenced by the need to accept that the human experience of living on Earth actually is the same thing as being Earth, or rather,

30. Lauren Ellis, "Behind the Scenes: Breeding Super Coral," accessed June 5, 2024, https://hilo.hawaii.edu/coralhealth/news/breeding-super-coral.php.

31. Hird, "Digesting Difference," 216.

as Rosi Braidotti puts it, "becoming-earth."[32] By taking up the task of redefining the human in relation to global and cosmic deep-time environments in terms of metabolism and "life after death," we need to reconsider our views on subjectivity and the ways in which we are part of inhuman forces and tentative material structures. Human bodily lives, viewed as metabolic processes and technologically augmented flows of matter and energy—call for recognition of what Kathryn Yusoff has termed "geologic life," postulating that we need to "see our ways of being as geological rather than biological per se … ."[33]

Today, humans equipped with powerful tools of synthetic biology and nanotechnology can rearticulate metabolism—both organic and nonorganic materiality—in ways not yet fully comprehended or predictable. They are shaping the environments of the future, which will likely sustain resilient, other-than-human life-forms. Inquiry into the sustainability of human metabolisms functioning amid global biochemical systems should not, however, be seen as a humanization of the environment, but rather as an assessment of the human impact on environments to come. Metabolic force generates novel forms of organization in an evolutionary process that allows metabolic systems—linked to specific organic and mineral species—to adapt to environmental conditions and, in their own unique way, participate in the global flux of matter. Hence, "a dynamic and sustainable notion of vitalist, self-organizing materiality"[34] seems necessary in dealing with the concept of metabolic force, leading to the dissolution of the living/nonliving divide. Yet, as Braidotti reminds us, "the planetary opens onto the cosmic in an immanent materialist dimension,"[35] not only because it is influenced by electromagnetic radiation from the Sun and outer space, but also because matter, both on Earth and elsewhere, is self-assembling and evolving, and thus always transforming itself. ∞

32. Rosi Braidotti, The *Posthuman* (Cambridge: Polity Press, 2013), 81.

33. Yusoff, "Geologic Life," 781.

34. Braidotti, *The Posthuman*, 82.

35. Ibid., 81.

Digital Content 02 Adam W. Brown, *The Great Work of the Metal Lover*, 2012.

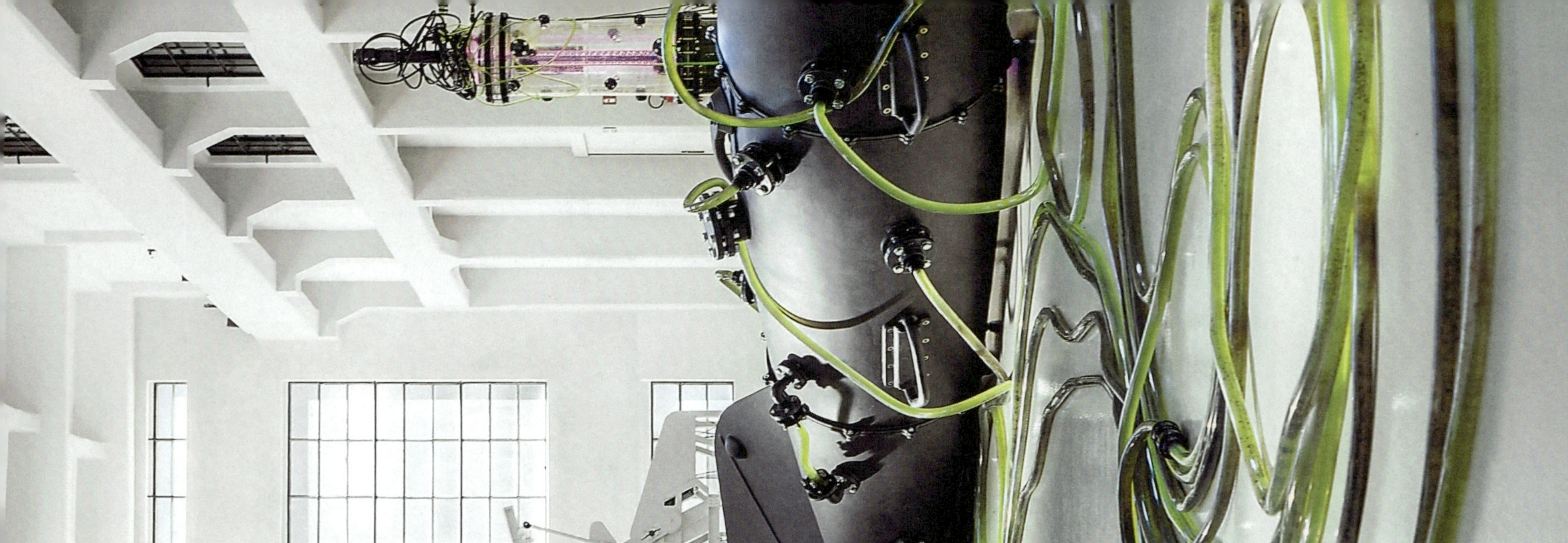

From Symbols to *Metabols*: Capacity for Synthesis in the Visual Arts

Anett Holzheid in Conversation with Thomas Feuerstein

Fig. 15 Thomas Feuerstein, *METABOLICA: MOBY DICK*, 2023, exhibition view *METABOLICA: MOBY DICK*, Museion/NOI, Bolzano. © Atelier Feuerstein

Poiesis: The Active Work

Anett Holzheid *Homo faber* (the human being as a maker or creator) is driven by a constant urge to effect change as well as by a logic that is technical and pragmatic. Humans' Promethean way of liÿe is not only evinced by the history of their artistic production, economic productivity, and skillful use of tools. Since the nineteenth century, scientific curiosity and inventiveness have produced insights into the biological processes of cell metabolism and into regulatory mechanisms and energy conversion in kinetic, thermal, chemical, and photonic reaction structures. This has resulted in far-reaching interventions into the material world. Justus Liebig gave an early account of the concept of *metabolism* as an incessant process of organic formation, transformation, and excretion: "Every movement, every manifestation of organic properties, and every organic action [is] conditioned by a change in the material of the body, and by the assumption of a new form by its constituents."[1]

Insights into the molecular composition of organic substances gave rise to an industrial metabolism that sought to break out of the closed system of natural recycling. In the late nineteenth century, in a kind of euphoria of liberation, fossil raw materials were synthesized and used to create everyday products such as paint, plastic, synthetic fibers, and medicines in the production facilities of the organic chemistry industry. With hindsight, we can now see that we have transformed the biological recycling system into one that is open and increasingly unstable. The accelerated and intensive advancement of chemical reactions is producing metabolic dysfunction on a global scale. Within these chemodiverse material flows, biodiversity finds itself in a critical relationship of dependence.[2]

Thomas, since the 1980s, you have been engaged in understanding digital, apparatus-based, and biochemical control cycles and processes from the perspective of the visual arts and in transferring them to the artistic workspace, where you furnish them

1. Justus von Liebig, *Organic Chemistry in its Application to Agriculture and Physiology*, ed. Lyon Playfair (London: Taylor & Walton, 1840), 369, translation modified.

2. Hermann Fischer, *Stoff-Wechsel: Auf dem Weg zu einer solaren Chemie für das 21. Jahrhundert* (Munich: Antje Kunstmann, 2012), 110.

with poietic productive resources. In the visual arts, the fields of exhibition architecture and artwork restoration are associated with building plans, molecular structures, and metabolism. You work primarily with molecular and machinic architectures and material syntheses—the artist's atelier has become a site of biotechnical fabrication.

Your works are interrelated in multiple ways and constitute a comprehensive oeuvre of processual art. Your *METABOLICA* series of works is now approaching completion. Let's take this dialogue as an opportunity to recapitulate and produce a kind of digest of your artistic understanding of sculptural experimental setups and metabolic processes in the visual arts and their links to other disciplines.

Your installation art projects, which perform processes of formation and decomposition, and bear eloquent titles such as *PROMETHEUS DELIVERED* (2017), are formally elaborated in aesthetic and functional terms. At the same time, their ongoing logic of change runs counter to any working assertion of completion.

Enclosed within the glass casings of the bioreactors, real liquid organic matter is presented and then driven to continuously transform. For me, your works are essentially manifestations of the capacity for synthesis. In the same way as you incorporate future-oriented questions from biology and biochemistry, and allow the organic and the inorganic to flow using circuit technologies, you also process ancient myths and combine fundamental motif cycles from the history of art and ideas. Your works oscillate within narrative systems as well as within discursive spheres of knowledge, and they fuel the dynamics of transfer between art, philosophy, literature, economics, technology, and the natural sciences. You fashion constellations of the material and the intellectual in lively contexts, and use them to mold dynamic time periods situated between ancient philosophy and the spaces of possibility afforded by science fiction. The processes of material autopoiesis—that is, self-reproduction—thereby

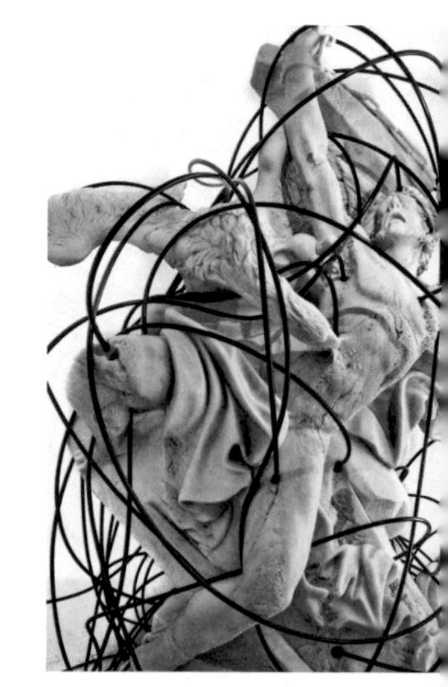

Fig. 16 Thomas Feuerstein, *PROMETHEUS DELIVERED*, 2017, exhibition view *PROMETHEUS DELIVERED*, Haus am Lützowplatz, Berlin. © Atelier Feuerstein

permeate aspects of the control systems of growth and decomposition. This stimulates the metabolism of the beholder's brain, excites their sensory apparatus, and presents a digestive semantic abundance akin to a feast of delights.

Every metabolism needs energy to drive it. Why do you find the visual art of interconnections and biocybernetic circuits, and the exploration of biogenic substances so compelling?

Thomas Feuerstein When I was a young artist, I asked myself what could change reality, the world, and therefore my everyday life. In the late 1980s, I was interested in information technology and biotechnology, and their utopian and dystopian promises. To this day, it is these spheres that comprise my core artistic inquiries because they are radically updating questions about being human, or rather the human condition, on a metaphysical level. For me, art was and is a medium through which to better understand the world epistemically, socially, and aesthetically. This curiosity persists to this day, allowing me to better understand the obsessions, dreams, and fears—from ancient myths to science fiction—that manifest in technologies.

While visual art usually operates on a symbolic level and reflects the world and society aesthetically or critically, living systems can be used to narrate contexts and stories on the level of living reality. This allows art to go further than the traditional framework of symbols, metaphors, and allegories, and to transcend ancestral iconic and linguistic domains. When it comes to my artistic method of conceptual narration, I like to interweave discursive, symbolic, and cultural-historical threads with real-life processes. Works thus become knots in networks that spin threads from history into the future. A knot is always an experimental way of arriving at new forms and possibilities. And from this knot emerge Ariadne's threads, leading art out of the bubble of representation and sometimes merely decorative critical commentary. The works contextualize themselves in these overlapping spheres of the natural sciences, economics, politics, and speculative utopias, but beyond that, I am fascinated by the metabolic characteristics of *Chlorella*. On the one hand, it is a mirror of our culture and the state of the world, and at the same time, it opens up possibilities to go beyond the symbolic functional contexts of art and to work in a processual way.

For me, processual sculptures are actants that put on a molecular, biochemical theater, engaging with symbols, icons, and above all with metabols. A sculpture, in this sense, is less object and more subject—that is, an acting, liyfing artifact that communicates and cooperates with other works. The artwork is the better artist in this sense, as it follows its own logic. This poietic logic constantly tempts me to come up with new questions, forms, and projects.

Projects and exhibitions are like Petri dishes incubated with questions and phenomena. In the medium and narrative of the artistic Petri dish, works begin to germinate, grow, and mutate. They condense, create connections, and literally and figuratively form biofilms that tell a story about everyday liyfe, liyfe itself, nature, and being human.

AH Considering the way you utilize photosynthesizing microorganisms and bacteria, your BioArt does not merely aim to represent the biological in art. You build technical environments for the natural processes of liyfe and formation, and you understand sculptural bodies from the perspective of their molecular substance and capacity for metamorphosis. You see connections between molecular combinations and multipartite organisms, allowing your work to diffuse ontologically between presentation, representation, and imagination. Social, model-like qualities begin to shimmer through from the biological formations presented. Your sculpture *GREEN HYDRA* (2021), for instance, stages the symbiotic relationship between *Chlorella* and hydras behind glass. Which conceptual difference do you see in this work compared to the two sculptures *MR. P.* and *MRS. D.* (2015)?

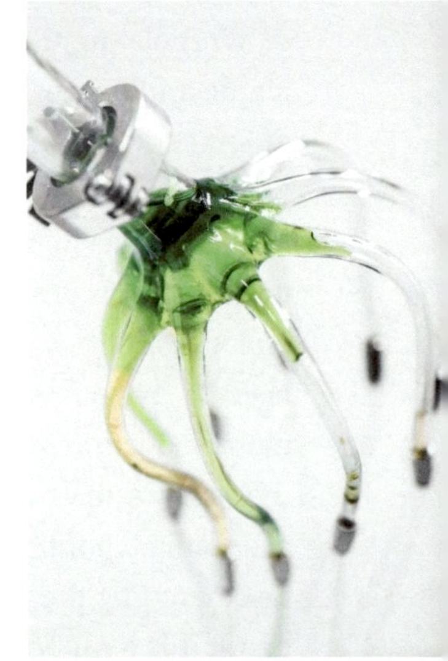

Fig. 17 Thomas Feuerstein, *GREEN HYDRA*, 2021. © Atelier Feuerstein

TF *GREEN HYDRA* is an important annotation to my works with *Chlorella*. *Chlorella* algae liyfe endosymbiontically in small freshwater

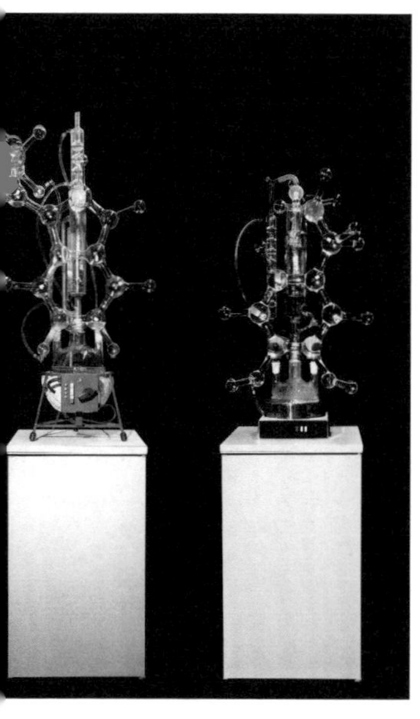

Fig. 18 Thomas Feuerstein, *MR. P.* (left) and *MRS. D.* (right), 2015. © Atelier Feuerstein

polyps (*Hydra viridissima*), allowing them to feed via photosynthesis. They are typical holobionts, that is, meta-organisms composed of different species that have evolved into a new functional life-form. Speculatively speaking, could this prototypically be an alternative to traditional economies and models of parasitic exploitation?

By contrast, *MR. P. & MRS. D.* are works that act as "workers," actants, or acting "subjects." They formed part of a production context and were the preliminary stage in the synthesis of the molecular sculpture *PSILAMIN* (2015), which was derived from the chemical synthesis of substances extracted from algae and fungi. *PSILAMIN*—for me, the smallest sculpture in the world—is a new molecule that, in crystallized form, remains almost invisible in the exhibition space. But as soon as you transfer the molecular sculpture from the exhibition space into your own body, into your bloodstream and into your brain, and "exhibit" it there, it changes your perception. *PSILAMIN* produces psychotropic effects, and the algae and mushrooms make us dream.

METABOLICA: Factory of Life

AH Your work series *METABOLICA* starts with a carbon cycle and includes a printer that squeezes out biomass like excrement for a sculpture. You are not interested in presenting the sculpture as an object. On the wall graphic for *METABOLICA*, you put Marcel Duchamp's urinal (*Fountain*) and Piero Manzoni's *Merda d'artista* (*Artist's Shit*) preserving jars in close proximity under the headline "HOW TO LOOK AT METABOLIC ART."

Alongside these wall graphics, you exhibit tin cans that contain a self-produced biopolymer powder, which you use as the material to 3D-print sculptures. All art is a source of energy and must be digested—that could be the premise here. What references from

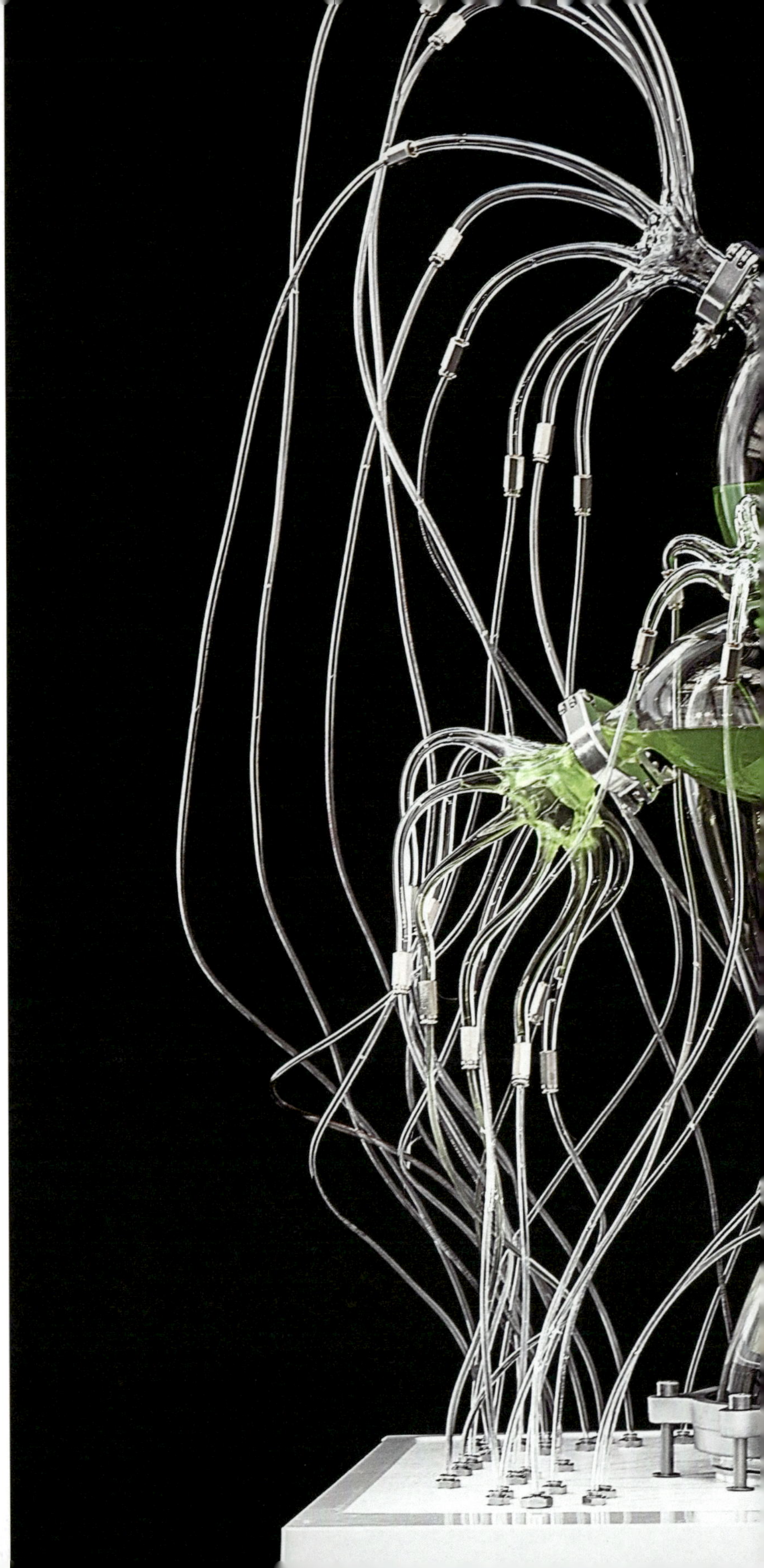

Fig. 19 Thomas Feuerstein, *GREEN HYDRA*, 2021. © Atelier Feuerstein

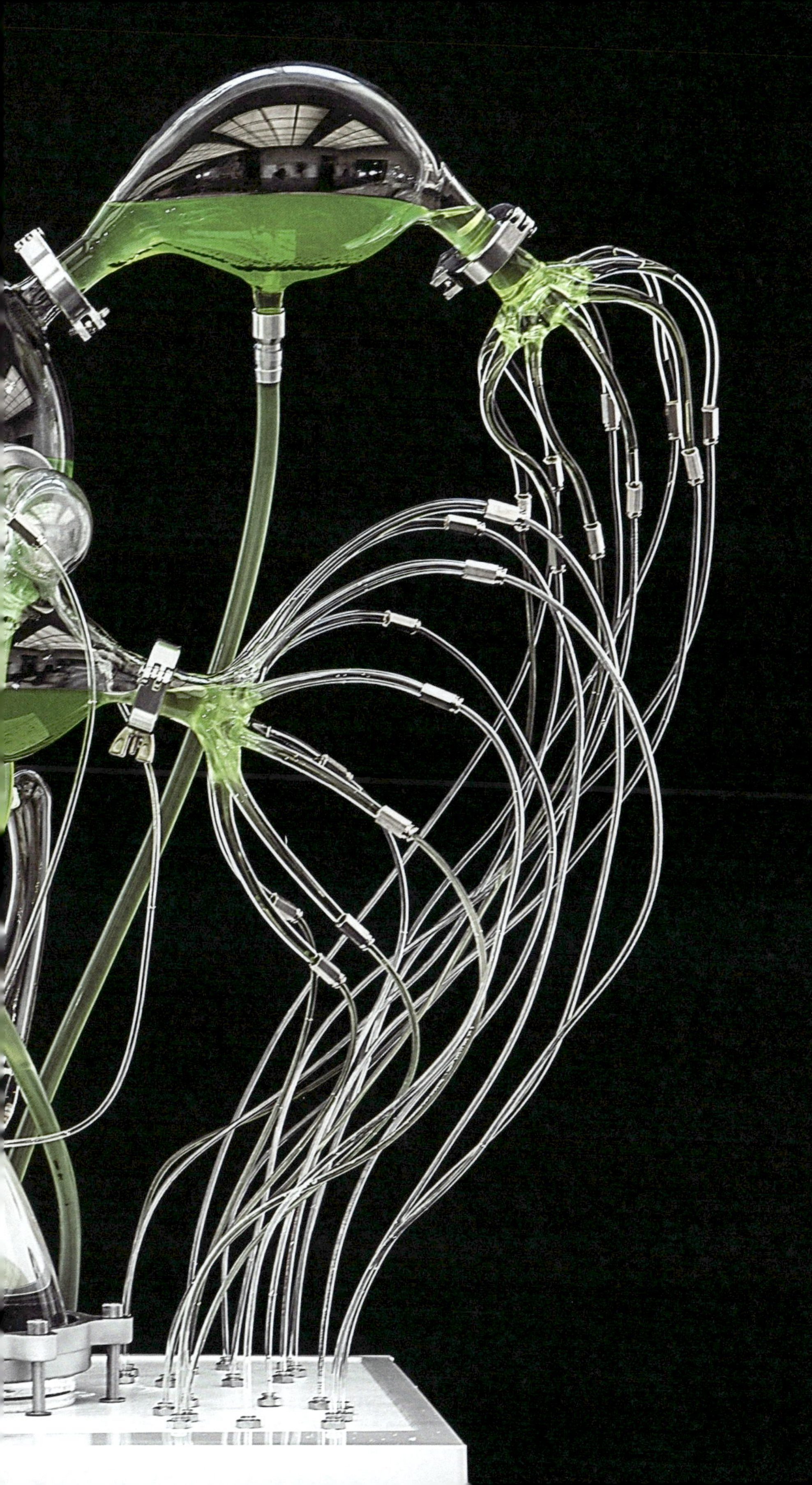

the history of metabolic art are important to you for your artistic *fabricca*/factory?

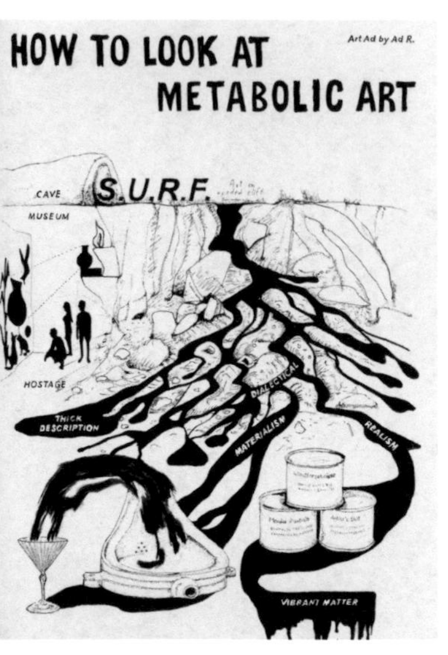

Fig. 20 Thomas Feuerstein, *How to Look at Metabolic Art*, 2022. © Atelier Feuerstein

TF Shortly before *Merda d'artista*, Manzoni created the multiple *Corpo d'aria* (*Body of Air*). It consisted of a balloon that is inflated by means of the mouthpiece provided. In both cases, the can and the balloon served as reservoirs for metabolic waste products. The body with its organs for digestion and respiration became an integral part of the works. Manzoni was influenced by Yves Klein, who had blue cocktails served at his 1958 exhibition *Le Vide* (*The Void*). In addition to gin and Cointreau, they contained methylene blue, which colored not only the glasses but also the bodily fluids of the cocktail-drinkers. Anyone who cried or sweated did so in blue. And anyone who used the urinal painted over Duchamp's *Fountain* excreted blue urine. These two works are important to me because they mark the beginning of metabolic art. Liÿe, the body, and its organs became an artistic medium and expanded painting and sculpture. Art transcended the symbolic and became subversive by getting under our skin or by emerging from what lay beneath the surface. Metabolic art promotes new alliances with reality, the biosphere, and being human. It not only represents but also processes and creates new connections with the world, which go beyond allegories and metaphors.

Thus, when I describe projects and exhibitions as *"fabbrica,"* this is not just a tribute to the artist's workshop of the Renaissance. In my exhibitions, I don't only exhibit; I also produce. This practice is expressed in the twofold meaning of the German term *Werk*, which either refers to a piece of art or to a production site: a factory. Our entire liÿe emerges industrially and cellularly from factories—from everyday items to our bodies. Factories constitute us physically, economically, and ecologically. For me, they are synonymous with liÿe and the labor activity of nature and culture, along with all the associated efforts and pleasures. Building a factory that grows and processes biopolymers is a perfect artistic choice for me. As an artist, I can criticize the world boldly and symbolically in my works or try to improve it. But the work remains only a commentary. What doesn't interest me as an artist is my own subjective state of mind or comments and opinions about it. I want to listen to the phenomena and problems of the world and get them to talk.

AH Listening to your environment-based motivation, sculptor and heat-energeticist Joseph Beuys comes to mind, who said that he drew the plasticity of his art from working with thoughts and language, and that, on the other hand, concepts for him arose from "feeling and willing."[3] In his "search for the real shape of things," Beuys achieved form through thought processes: "I always start with what I can manage, but I am only interested in making a new sculpture if that new sculpture also contains a question for me."[4] The starting point for the new is "entirely undifferentiated, a chaotic flow."[5] He saw human will, as well as the human digestive system, as "the starting point for the forces that make use of the chaotic. In other words, as freely flowing, chaotically flowing material."[6] Beuys relied on introspection; you focus on the biosphere. Your *METABOLICA* project consists of multiple material flows that have process converters with agent status. What actuates *METABOLICA?*

TF In the current *METABOLICA fabbrica*, sculptures and organisms become collaborators and actors in a process that utilizes algae and bacteria to produce material for new sculptures. *METABOLICA* prototypically sets a cycle in motion in which bacteria provide the

3. Joseph Beuys, "Talking About One's Own Country: Germany; Lecture Given at the Kammerspiele, Munich, 20th Nov. 1985"; reprinted in *In Memoriam Joseph Beuys: Obituaries, Essays, Speeches,* trans. Timothy Nevill, ed. Wilfried Wiegand (Bonn: Inter Nationes, 1986), 37. Beuys continues: "The precondition for a successful sculpture was thus that an inner form first came into being in thought and understanding which could then be expressed in the shape of the material used in the work." Ibid.

4. Rolf-Gunter Dienst, "Joseph Beuys: Interview," in *Noch Kunst? Neuestes aus deutschen Ateliers* (Düsseldorf: Droste, 1970), 47 (transcript of an interview from December 1969).

5. Ibid.

6. Joseph Beuys, "Werkstattgespräch, 6 June 1969," interview by Hanno Reuther, WDR, July 1, 1969; published in Kunstmuseum Basel, *Joseph Beuys: Werke aus der Sammlung Karl Ströher* (Basel: Kunstmuseum Basel, 1969), 41.

material for sculptures—in the same way that the marble quarries in Carrara have done since Ancient Rome—and at the same time they function as sculptors' chisels by breaking down the sculptures again and changing their shape. *METABOLICA* is substance change in a biological and economic as well as an artistic sense. The works tell of the end of petromodernity, of a historic *metabolē*, as a changeover from petrochemistry to biochemistry.

AH In Greek, the term *metabolē* refers to a profound transformation, a complete reversal, like water to fire, alive to dead. It seems to me that your understanding of metabolism does not align with that of ancient Greek philosophy and its notion of sudden changes. Rather, your work seems to be the driving energy behind thought processes that seek to initiate a movement of fundamental transformation. You use the art space to sculpturally formulate a statement that advocates an ecological and economic rethink. With *METABOLICA*, you present a new aesthetic whose beauty and elegance also consist in imagining metabolism as something that is efficient, economical, targeted, effective, and low-waste, just as it is in natural chemical biosynthesis.

Let's take a closer look at your works. What does the submarine signify that appears to have sunken to the floor of the exhibition space, with which you open the first chapter of *METABOLICA*? As an object from biotech fiction that has been interwoven and is still unidentified, does it embody the salient point of the *METABOLICA* project? A chimera, a *HYDRA* you call it, a unique original, a stranded, slender baleen whale? Or is it not seeking to be anything more than a symbolization of Aristotle's *hyle*? With its self-contained, dark-gray polished surface, at first sight it looks strangely velvety and softly matte when viewed from afar. Its weight is difficult to gauge: hollow and light from a distance, much more technical and massive when we get closer to the sculpture, yet with consistently pleasing proportions. The algae tubes grow out of it like tentacles, which, as they have a similar

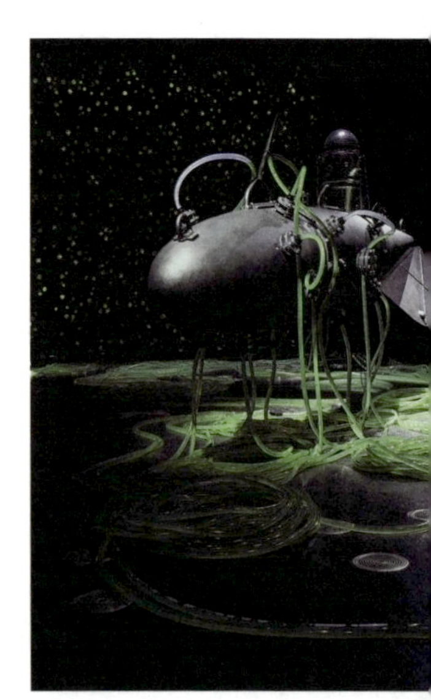

Fig. 21 Thomas Feuerstein, *HYDRA*, 2020, exhibition view *gREen: Sampling Color*, Muffatwerk, Munich. © Atelier Feuerstein

function to the earlier *MANNA* sculptures, transform the exhibition space into a "photosyntheticon." After meandering for *n* kilometers per meal, they are finally led back into the body of the whale.

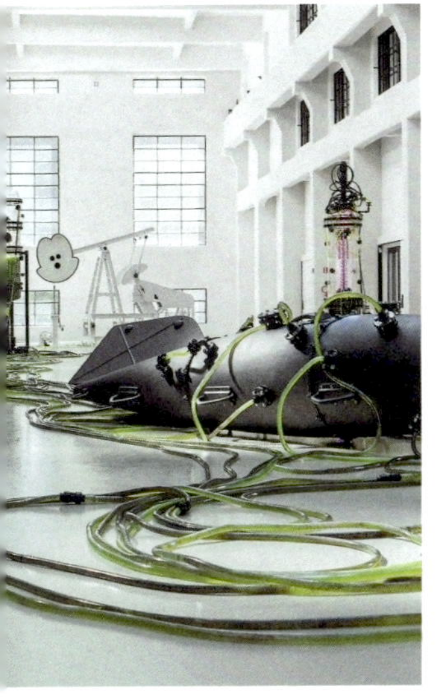

Fig. 22 Thomas Feuerstein, *METABOLICA: MOBY DICK*, 2023, exhibition view *METABOLICA: MOBY DICK*, Museion/NOI, Bolzano. © Atelier Feuerstein

TF *METABOLICA* performs carbon cycles as the basis of life and tells a story of cultural change as material change. It begins with the work HYDRA, which uses floating algae as a carbon sink. As in nature, the cycle begins with photosynthesis: with light, water, and carbon dioxide. The algae meander through a system of transparent tubes; they absorb light, reproduce, and store energy, which becomes the carbon source for bacteria in the bioreactor sculptures *MRS. MOL* and *MR. MOL*.

I'm delighted that you think of *hyle* in this context! The original Aristotelian material *hyle* is the Greek word for wood, the "plastic" of antiquity, which was used to make all kinds of everyday objects, including houses. Like the biodegradable plastic PHB produced by *METABOLICA*, wood is a renewable material that can be shaped into different forms using *téchnē*. While wood can only be shaped through processing, that is, brought into a certain form, the concept of technology in biochemistry and materials science goes deeper today, enabling material properties to be programmed. For example, using the appropriate processes and additives, PHB can be foamed, stretched into a film, injection-molded, or 3D-printed. It can be hard and brittle or flexible and soft. Aristotle illustrated the relationship between matter and form as well as the technique used to convey them with the example of the sculptor Polykleitos and his sculptures. Using art and practical skills, the sculptor shapes matter and gives it form. One could also apply such a relationship between matter and form to painting and other art genres, but I think that artistic processes today are less causal and based more on interaction. Matter is not passive, "feminine," and just waiting for the "masculine" idea to give it form; matter is also always "vibrant" in the sense intended by political theorist and philosopher Jane Bennett, for example. My concept of sculpture therefore starts with matter and not with form. Substance theory and the concept of *hyle* are actualized both scientifically and artistically, for atoms and molecules are inconceivable without form and structure. *METABOLICA* as a biochemical, molecular factory starts with glucose, which can be described as the primordial substance or *hyle* of liyfe, because it is the universal fuel for all cells. *HYDRA* produces *hyle* in the form of glucose, which is metabolized into fatty acids and then converted into PHB by bacteria. This means that sculpture and art today are thinking about form above and beyond any external shape, which is expanding aesthetics and rendering deeper processes perceptible over and above beautiful surfaces.

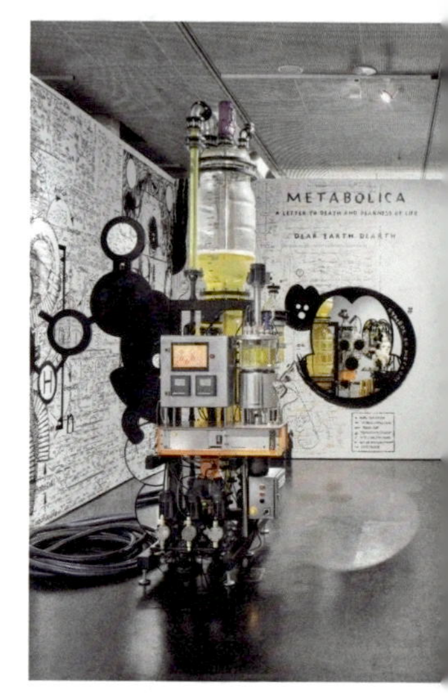

Fig. 23 Thomas Feuerstein, *MS. MOL*, 2023, exhibition view *Renaissance 3.0*, ZKM, Karlsruhe. © Tobias Wootton

AH *METABOLICA* refers to Aristotle's cosmology in several respects. In Aristotle, on whose writings on the philosophy of nature and art your works are oriented, *metabolē* refers to either a change to things or to something underlying that change (such as matter), which both take a certain amount of time.

Your neologism *METABOLICA* also appears to allude to Aristotle's compendium of texts, *Metaphysics*, from which you have already quoted. So, with *METABOLICA*, are you taking a holistic ecological, biotech-humanistic perspective, using your work—i.e., a comprehensive narrative—to present the fundamental and, for humans, still astonishing processes of liyfe within the context of cycles of becoming and passing away?

TF In Aristotle's *Metaphysics*, the concept of permanent change as a prerequisite for becoming and passing away is impressive. It is about liyfe and death. The Aristotelian concept of *hyle* already contains the idea of biological and economic cycles, which I describe as "ourobocracy." Ourobocracy extends beyond the rule of humans and includes the material and energy cycles of the entire biosphere. A brief example: in the case of whaling, the belief that the near-extinction of whales would lead to a massive increase in plankton was specious. In the Southern Ocean alone, whales once filtered and consumed more krill than is currently extracted by industrial fishing worldwide. However, metabolic cycles are complex and led to the Antarctic paradox: when the hungry whales disappeared, so too did the plankton. Suddenly, there was no fertilizer in the form of whale feces left floating on the ocean's surface. The cycle was interrupted because those nutrients sank to the seafloor with the dead plankton, and the metabolic productivity of the ocean decreased dramatically. Melville's white whale and the black fossil oil derived from sedimented plankton are like two sides of the same coin. Metabolically, hope resides in the in-between, between liẏe whales and fossilized biomass, in the liẏing plankton. Neither whale oil nor the black juice of plankton corpses can satisfy the hunger for energy and raw materials in the long term. *HYDRA* swims precisely in this in-between: in the interests of the biosphere as a molecular metaphor of translation into alternatives or—ecologically speaking—into possibilities that have no alternatives.

Submarines and spaceships are themselves bioreactors for people to leave their ancestral biosphere. As a metaphor, *HYDRA* presents us with the choice of continuing petromodernity in artificial spheres or ending it and taking it forward into natural spheres. This is where Aristotle comes in again: for him, matter is shaped culturally via technology and, of course, by the soul. Liẏing matter, including our bodies, is informed by the soul, and maybe in this sense biotechnology could become an ensouled *téchnē*.

AH In clear distinction to mechanistic concepts, you locate ensouled biotechnology in the future. This is reminiscent not least of Patrick Geddes's "neotechnics" and his now somewhat forgotten book *Biology* (1925). According to this Scottish biosociologist and urban planner, future biotechnology should also include psychotechnology: "true Biotechnics has as far as may be—and thus above all in human liẏe and education—to be also Psycho-technics. Industry

(Technics) has to be a 'good job'; and it thus becomes eutechnics, as were the crafts as well as arts of old."[7]

Let's be a little more specific about the "good job" in your change-oriented *fabbrica*. In an early text of yours on translocation, you state that it is the task of art "to design methods of aggregation and increased complexity that are capable of setting in train processes of perception both inside and outside places and their boundaries."[8] In your exhibition spaces, you design narrative architectures with an instrumentarium consisting of wall-filling graphics, voluminous bioreactors, pumps, 3D prints, mirrors, and much more besides. You scale microorganismic processes to fill the space and include the human visitor graphically as a baseline within a reorientation structure of changing media and material circulation processes as well as communicative, associative connections. This strategy becomes apparent at the latest when we take a closer look at your installation *FROM HAND TO MOUTH* (2023).

You not only foreground the premier tool of the Promethean human but also submit the artificial model hand to a process of slow bacterial decomposition. In *FROM HAND TO MOUTH*, this oversized body part is actually a 1:1 PHB replica of Michelangelo's monumental *David*. This entropic staging and defamiliarization strategy triggers us to reconsider figures of thought such as the political function of sculpture in the Florentine Republic, the idea of the human being that has developed since the Renaissance, or the question of power, forces, and laws of energy. What contribution does the installation *FROM HAND TO MOUTH* make within the *METABOLICA* project? How does it integrate as a logical part of the process?

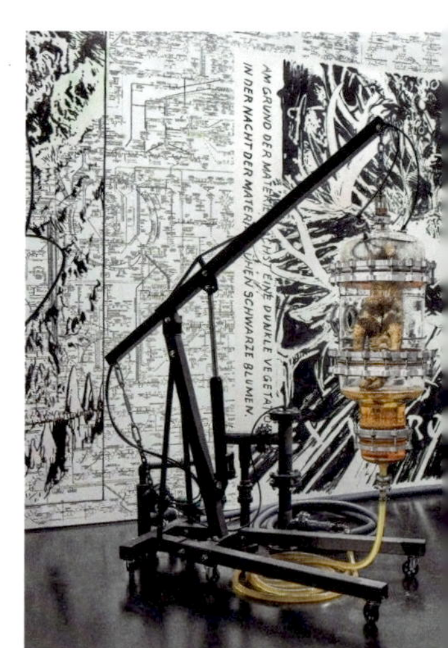

Fig. 24 Thomas Feuerstein, *FROM HAND TO MOUTH*, 2023, exhibition view *Renaissance 3.0*, ZKM, Karlsruhe. © Tobias Wootton

7. Patrick Geddes and J. Arthur Thomson, *Biology* (London: Williams & Norgate, 1925), 246.

8. Thomas Feuerstein, "Der Künstler als Translokateur," in *Translokation: Der ver-rückte Ort; Kunst zwischen Architektur*, ed. Marc Mer, Thomas Feuerstein, and Klaus Strickner (Vienna: Triton, 1994), 119.

TF Your questions address central aspects of *METABOLICA*. What are "eutechnologies," that is, technologies in industry and in work that make us happy? Likely they are those that do not alienate us psychologically or ecologically. Production that aims at an unlimited increase in the value of capital inevitably alienates us from natural cycles and ultimately leads to an "irreparable rift"[9] between humans and nature. Karl Marx, inspired by Justus von Liebig, recognized this very early on and argued for a fundamental change in the sphere of labor in order to patch up this rift. It is no longer possible to separate the sphere of human labor from the bio-, geo-, hydro-, or atmo-sphere: a reformulation is required of what we have hitherto laid claim to as human culture in the sense of sovereignty over natural metabolic cycles.

The motif of the hand in *FROM HAND TO MOUTH* stands for work and technology. But we don't live literally from hand to mouth. The hand symbolizes the mediating organ between culture and nature, and deliberately cites Michelangelo's *David* as a symbol of humanism and individualism. However, unlike David, it does not hold the stone ready for its sling, but bathes in a bacterial solution. It feeds the bacteria that have produced the material it is made of. It does not take, it gives. At the same time, it forms both the end point and, in a sense, the Renaissance, as in the rebirth of something new. With *HAND TO MOUTH*, the metabolic process of *METABOLICA* draws to a biological and artistic close.

AH Now our consideration of *METABOLICA* has actually led from *HYDRA*—the process-initiating starting point with reference to carbon change—to the sculpture *FROM HAND TO MOUTH*, which marks the end of the narrative. Atmospheric physicist and climatologist Hans J. Schellnhuber describes carbon (C for the Latin *carboneum*) as the basic material of creation, if one believes in a world creator. Schellnhuber ties the significant similarities between the bodies of Marilyn Monroe, Michelangelo's *David*, and the Dresden Green Diamond to this element, and reminds us that all known life-forms and stages of development, as you have already mentioned, originated in

9. Karl Marx, *Capital: A Critique of Political Economy*, vol. 3 (New York: Penguin Books, 1991), 949.

organic carbon compounds and are dominated by them. The bonding ability of the carbon atom allows for enormous chain molecules to form, leading to "highly complex planar and spatial structures with fascinating properties … . The sculptor's dream material owes its existence to carbon."[10] When Schellnhuber attributes the attractiveness of Carrara marble to its unique silky sheen, which imitates flawless human skin, this leads me to your *METABOLICA* sculpture *AHEAD* (2023), a 3D-printed replica of *David's* head printed using biopolymer. "Mother-of-pearl" and "silky sheen" are also the perfect attributes to describe the effect of this material. However, in my opinion, the seductive qualities of your biopolymer material in the eyes of the viewer no longer derive solely from the analogy to human corporeality. Rather, the rebirth of form in a new and seemingly pure material appearance directly raises the possibility of new materials and the ecological viscosity of their decomposition and degradation processes, which require further exploration.

But haven't we come too far too quickly? In order to better understand the artistic poiesis of *METABOLICA*, would you like to explain the biochemical metabolic processes once again, from chapter to chapter in context, without any interruptions from my side?

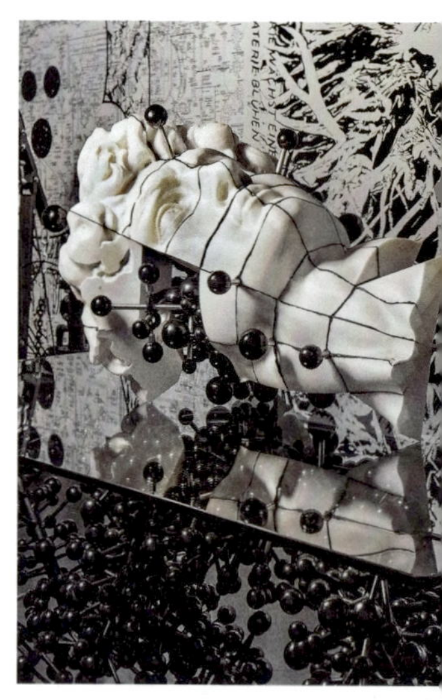

Fig. 25 Thomas Feuerstein, *AHEAD*, 2023, exhibition view *Renaissance 3.0*, ZKM, Karlsruhe. © Tobias Wootton

TF *METABOLICA* is a molecular, novel-like narrative in five chapters. In the first chapter, green algae reproduce by means of light and photosynthesis in *HYDRA*, but they do not yet contain any acids. In a further cycle, the algae in the glass sculpture *FATTY FANTASY* are put on a special diet: the medium, water, does not contain fertilizer or nitrogen, which means the cells cannot form amino acids or proteins, and therefore cannot reproduce. Their metabolism changes, and the cells produce fatty acids. Just like humans

10. Hans J. Schellnhuber, *Selbstverbrennung: Die fatale Dreiecksbeziehung zwischen Klima, Mensch, und Kohlenstoff* (Munich: Bertelsmann, 2015), 181.

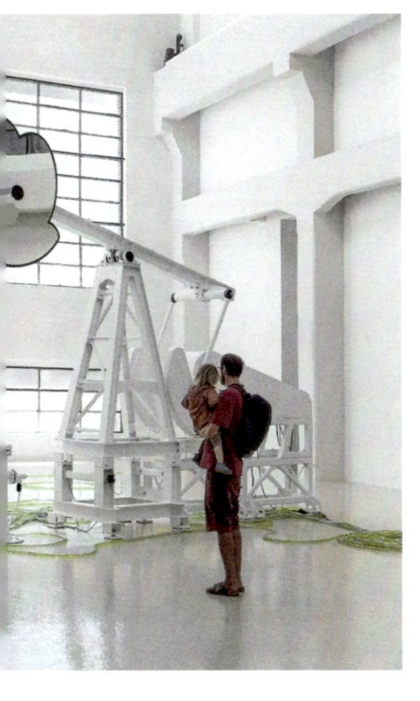

Fig. 26 Thomas Feuerstein, *MOBY DICK*, 2023, exhibition view *METABOLICA: MOBY DICK*, Museion/NOI, Bolzano. © Luca Guadagnini

increase their belly fat while consuming peanuts and lounging on the couch, the algae store energy.

The *FATTY FANTASY* cycle is powered by *MOBY DICK*, a deep or horsehead pump commonly used in oil production. The cycle could just as easily be accomplished using a good garden pump, but the shape of MOBY DICK is iconic and symbolic of old, outdated techniques of exploiting nature, which are repurposed in *METABOLICA*. *MOBY DICK* extracts not old, rotted plankton in the form of fossil oil, but photosynthetically renewable plankton.

In the second chapter, bacteria (including *Cupriavidus necator*) metabolize the fatty acids of the algae into polyhydroxybutyrate, PHB. Biosynthesis takes place in the two reactor sculptures *MRS. MOL* and *MR. MOL*, and is comparable to the processes in *HYDRA* and *FATTY FANTASY*. If nitrogen is present, the cells reproduce—this occurs in the *MRS. MOL* sculpture; if it is absent, the cells form PHB as a storage substance—this occurs in the *MR. MOL* sculpture.

In the third chapter, the PHB accumulated in the cells is extracted in the sculpture *REFINERY*. When dried, the result is a fine, white powder, which is then processed into new sculptures by a 3D printer in the fourth chapter. PHB is a thermoplastic polyester with a melting temperature of around 170 degrees Celsius. It can be used in conventional injection-molding machines and, in addition to using it for 3D printing, I also melt it in cooking pots, which is risky—and I don't recommend you try it at home. However, the hot melt can be poured into molds or molded into sculptures.

For me, the special 3D printer in the fourth chapter is a sculpture machine, which I named *ANAKEL* in reference to the Oracle of Delphi. The Pythia used to sit on a tripod over a chasm in the rock and spoke her prophecies. *ANAKEL* is a Delta 3D printer, and its extruder nozzle sits on an inverted tripod. *ANAKEL* therefore does not speak orally, but anally. 3D printers press hot thin sausages out of their rosettes, and *ANAKEL* uses this principle for the metabolic product PHB, that is, for its sculptural "prophecies." I see this as the artistic outcome of *METABOLICA's* digestion process. During the printing process, the printing plate slowly sinks into a reactor glass filled with water, where the bacteria begin to feed on the objects. This same process continues in the sculpture *FROM HAND TO MOUTH*, which is part of the fifth chapter.

In the fifth chapter, printed, cast, and modeled sculptures come together to form a *WHOLE DEARTH CATALOG*. Some are undigested, others are partially metabolized, and others are undergoing bacterial degradation *in vitro*, as in *FROM HAND TO MOUTH*. The sculptures metamorphose references from the history of sculpture, whaling, and petromodernity. They function symbolically and metabolically as transitional objects and stand for a historic, cultural, and industrial transformation. Over and above the iconographic references, and going beyond their individual forms, the sculptures speak through the poiesis of the material, which creates a distinctive aesthetic. Meaning in the sense of the signifier and the material carrier of meaning in the sense of the signified fuse and decompose again at the same time. To me, this is the point where the biotechnological endeavor reveals itself to be an artistic process and creates a unique quality of visual art that cannot be reduced to the surfaces of images or forms, or a conceptual literary narrative. The materiality and the processes associated with it that have always been intrinsic to visual art culminate in an artistic manifestation. This reveals the potential of contemporary art to condense the phenomena of life in order to fertilize our desires, hopes, fears, and imaginations with real-life processes in nature, in our environment, and in our culture.

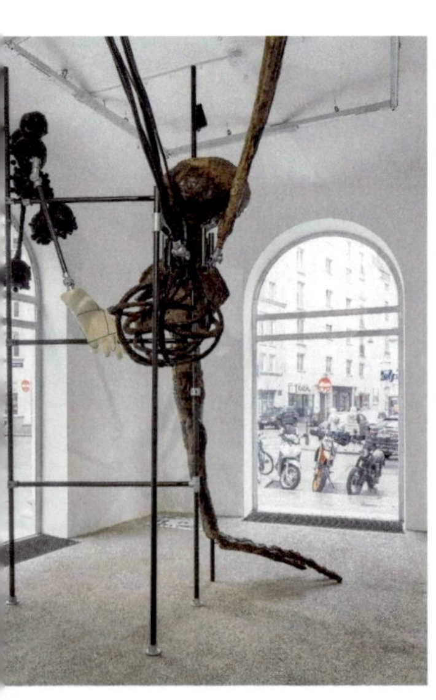

Fig. 27 Thomas Feuerstein, *WHOLE DEARTH CATALOG*, exhibition view *WHOLE DEARTH CATALOG & GOOD ROTTEN GOODS*, Gallery Elisabeth & Klaus Thoman, Vienna. © Gallery Elisabeth & Klaus Thoman

The title *WHOLE DEARTH CATALOG* illustrates this by playing on the title of the legendary *Whole Earth Catalog*, a counterculture magazine published by Stewart Brand from 1968 to 1972. The Earth, its biosphere, and atmosphere, its wealth of resources and biodiversity, are becoming scarce: *Earth* is becoming a *dearth*. And *dearth*, through its etymological connection to *dear* and *dearness* (in the sense of precious), testifies to a longing born of scarcity for what is most missed and needed, be it friendship or love, food, energy, or information, justice, happiness, or knowledge, resources, prosperity, or the future. As an open, unfinished work, the *WHOLE DEARTH CATALOG* brings together sculptures whose material composition transforms the non-finiteness of sculpture into a metabolic infiniteness. The bacteria that produce the sculptural material PHB can change the shape of the biological sculpture or break it down completely in order to generate new material. They are sculptural collaborators, and the resulting sculptures have both a symbolic and a metabolic form. They are symbol and "metabol."

AH Indeed, your take on PHB turns it into quite an extraordinary symbolic and metabolic material. It is a biopolymer (*biologischer Kunststoff*) that occurs in nature, so you rightly refer to it as a *Naturstoff*. Being produced in your artistic factory however, PHB becomes *Kunststoff* in the sense of becoming an artistic material as well, which you use to synthesize important issues: the social metabolism, the renewal of the polity, the inclusion of digestive processes, and the role of the intestine as a second brain. Would you like to elaborate a bit more on what the chances are of moving away from our current one-way energy economy to a better cyclical order in which our digital metabolism could also work well?

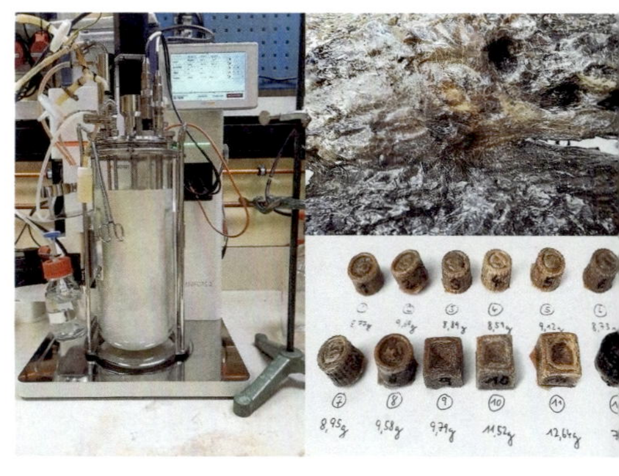

Fig. 28 Degradation tests of PHB samples for *METABOLICA*, Department of Microbiology, University of Innsbruck. © Atelier Feuerstein

TF Regardless of the great potential of the material PHB to replace petrochemical plastics such as polypropylene and industrial products made from it, from car bumpers to plastic foil, *WHOLE DEARTH CATALOG* demonstrates how a universal changeover from a capital-driven plutocracy to an "ourobocracy" could take place. Unlike modernist linear production chains that lead from assembly line to consumption to landfill, the Ouroboros will become the patron of a metabolically and cyclically organized polity. "Ourobocracy" stands for an integrated and iterative process of transforming matter, all goods, and the way we consume. So far, the sovereign of a state has usually been associated with a human brain that governs the body of the people, as in Thomas Hobbes's *Leviathan*. In the ourobocracy, this would instead be the belly, the gut and its microbiome, through which the world flows as food and which incorporates all spheres of nature. The ourobocracy will raise questions about fundamental rights, the state's monopoly on the legal use of force, the ruling class, and capital. What is the status of a state? What is economic growth? To this day, economics is primarily based on metabolic end products such as fossil fuels or goods with planned obsolescence. I do not see art, on the other hand, as a metabolic end product, but always as part of a historical, intellectual, emotional, and aesthetic digestive process, which in *METABOLICA* has a socioeconomic, ecological, and political effect through the actual material of the series, which is both a symbol and a "metabol."

AH What significance do you attach to PHB-producing microorganisms?

TF PHB-producing bacteria interest me as model cultural organisms. On an artistic meta level, they act as protagonists in a narrative like that of a parable, and function exemplarily as allegory or metaphor. Yet as a real metabolism, they expand the field of art and burst out of the confines of symbolic commentaries and aesthetic illustration. In terms of biochemistry and enzymes, they function as molecular ateliers and factories in which tools for change are available that have developed evolutionarily. I owe the opportunity I have had to explore and make productive use of these possibilities to my collaborations with scientists who, on the one hand, have provided the technical equipment involved in these processes that is necessary for my artistic work and who, on the other hand, have developed bioeconomic processes that go beyond the boundaries of art. I was able to attract microbiologists for *METABOLICA*, not because of their enthusiasm for art, but because they are interested in fecal matter in sewage treatment plants. Wastewater contains a variety of carbon sources and energy, and the aim is to use them for the production of PHB in the near future, in addition to producing biogas. I see this as a concrete link to your question about social relevance or social metabolism. Why are a variety of objects, especially those that unintentionally end up in the environment, not being produced from PHB instead of from petrochemical, nonbiodegradable plastics? Microplastic contaminates ecosystems, including our own bodies, which could be avoided by using PHB. Food packaging, fishing nets made from PHB, and so on, could be the start.

Narratives of Art

AH In the midst of your receptacles that communicate materially and metabolically with each other via media, the most diverse discourses and particles—molecules, cells, letters—all become participants. You use them to cultivate stories of a possible future—you conceptualize the future. In your earlier large-scale projects like *PANCREAS, PSYCHOPROSA*, and *PROMETHEUS DELIVERED*, you developed this method of "conceptual narration." Why is conceptual narration a compelling artistic "eutechnique" and one of the pivotal methods in your extensive program to metabolize art? Do you perhaps see it as a way of offering conceptual art, which has now become somewhat arid, a lively, future-proof environment? You have already touched on the beginnings of metabolic art.

TF From its beginnings to the present day, I regard conceptual art as diverse and differentiated. Works by Robert Smithson and Joseph Kosuth, for example, take very different methodological approaches. Conceptual art is commonly understood as an immaterial discipline in which language and theory predominate. For me, it goes beyond that and addresses and includes materialities like no art before it. *7000 Oaks* by Joseph Beuys, for example, can be understood as a green manifesto, but also as the largest-ever wooden sculpture. Its authorship has expanded to include the inhabitants of the city, institutions, commercial enterprises, and, last but not least, the trees themselves. This aspect, which brings different actors, political and ecological realities, social and biological processes into play, has created a new artistic narrative.

In addition to extended authorship, real-world references, and processuality, conceptual narratives manifest for me as literary/linguistic, iconic/pictorial, and material/molecular all at once. Unlike the idea of the *Gesamtkunstwerk*, conceptual narratives do not seek to fuse individual works and art genres into a totality, but instead aim to form networks. This alters the narrative style and makes specific qualities of visual art aesthetically productive. Literature and film are great, effective media, but they can only tell stories as a linear progression and represent their narrative in language and images. Visual art can directly incorporate materials, processes, and entire biotopes, give them a voice, and turn them into collaborators in a work. In many respects, conceptual narratives are polyphonic narratives that also open up to phenomena that stand outside the power of linguistic and pictorial symbols as well as cultural experiences and logic. Visual art is predestined to transcend the boundaries of the symbolic and to tell stories metabolically, though metabolism is not limited to biological metabolic processes. Metabolism includes cultural metabolic processes in industry and consumption as well as in information technologies. In particular, the sucking up and digesting of data streams, which is becoming increasingly important in the field of machine learning, is turning the internet into a gut and establishing an enteric system for a digital metabolism.

AH Jean-François Lyotard's postulated end to the grand narratives also fell in the 1990s. With all the dynamics of pulsating prompts and posts that exceed the human capacity for reception, hasn't the time come once more for a new emphasis on texts devoted to contexts? How do you view the relationship of your "conceptual narration" to the new materialism and actor-network theory? Who narrates what? What would be the additional artistic, aesthetic, or epistemic value of narratives by nonhuman actors?

TF Storytelling in literary traditions is human. When nonhuman actors speak, such as gods in myths, animals and plants in fables, or aliens in science fiction novels, they do so vicariously and allegorically. The Other functions as a rhetorical figure of the same masked as personification. Even animist stories and scientific papers that tell of natural forces appeal to the human imagination, to human spirituality or cognition. But how can all the nonhuman entities that determine our existence, such as bacteria and fungi, the ocean and forest, the climate, the microbiomes of the soil or the gut, be made to speak? Do we have to learn to tell stories in foreign tongues, and does this require futuristic techniques of xenoglossy?

If there is no getting away from language, with or without the linguistic principle of relativity, the question for art becomes whether narrating is actually not only thinking but also acting. This presumes that artworks have a power to act that transforms passive objects into active subjects. Works of art thus go beyond being carriers of meaning and become carriers of action that create a logic of their own as a form and an aesthetics that is detached from authorship and is independent of stylistic intentions, the intentions of the artists, and the interpretations of the viewers. Literature has hallucinated the vitalization of objects since antiquity at the latest, albeit predominantly as something anthropocentric or demonic. In Ovid's *Metamorphoses,* the figure of Pygmalion already etymologically contains the material of Goethe's *Faust* through *pygmē* (Greek for fist), and in Romanticism, the independent life of literary works brought forth the modern horror genre, for example, Edgar Allan Poe's *Life in Death* or Oscar Wilde's *The Picture of Dorian Gray*. In contrast to the bringing of statues and images to life in literature and their psychological reflection of human emotional worlds, post-humanism, new materialism, object-oriented and relational ontology, and actor-network theory work against subject-centeredness. It is not people who are called upon to tell stories, but things and processes. Descartes's substance dualism as a distinct boundary between thinking subjects (*res cogitans*) and inanimate objects in the environment (*res extensa*) is becoming blurred. This is creating space for stories that do without metaphors and questions of meaningfulness. Nonhuman entities are becoming the protagonists in new narrative styles that are engaging with the design of technical, social, and ecological spheres. Paradigmatic of these narratives is that they focus on actions that make materials, processes, and biological organisms speak. The performative expands the narrative in that it is not only the artists who speak through their works but also object-immanent materialities.

When Robert Smithson speaks of “collaborating with entropy”[11] and lets viscous liquid flow from a barrel over the ground, he is addressing the intrinsic nature of objects. It is not the artist who is the creative “dictator” of the form, it is the material, gravity, the ground, the ambient temperature, and so on—all these factors become active collaborators in Smithson’s work. The artwork does not remain in an immutable state like a sculpture made from bronze or stone; in fact, the artwork begins to empower itself. This results in a poiesis of art that self-produces works to a not insignificant extent. Artists utilize real-liyfe processes and make explicit the old insight that tools and materials, media and production conditions, are a constitutive part of art. In contemporary art practices, works are thus increasingly becoming “symposia” in which human and non-human guests come together to explore aesthetic areas of activity. Materials, chemical reactions, liyfing organisms, algorithms, and data streams are becoming actants that are expanding authorship and trying out new aesthetics.

AH You are busy compiling a sculptural vocabulary that promotes metabolic insights. In other words, the conceptual narrative you insert into your work is progressive proof of activity. Your artistic self-image is sustained by an appeal for change in current ecological and economic conditions in such a way that activity no longer needs to be referred to as revolution in your metabolic cosmology. Holding onto your image of art as a Petri dish as a final thought, I would like to reinforce the art-catalyzing and future-oriented power of your process-sensitive work with a footnote from the era of the first ecology movement. This original short quote by Beuys on the conditions for success demonstrates how the materials of intellectual liyfe change from one generation of artists to the next: “ ... if the revolution does not first take place within humans, every external

11. Robert Smithson, *The Collected Writings*, ed. Jack Flam (Oakland: University of California Press, 1996), 256.

revolution will fail. Humanity must conquer the interior space, just as astronauts conquer outer space. In any case, art always speaks to the individual, free, creative person."[12]

TF That's the perfect ending for a new beginning. ∞

Digital Content 03 Thomas Feuerstein, *GREEN HYDRA*, 2021.

12. Joseph Beuys, telephone conversation with Armin Halstenberg; see Armin Halstenberg, "'Jeder Mensch ist ein Künstler,'" *Kölner Stadt-Anzeiger*, June 14, 1968, 36.

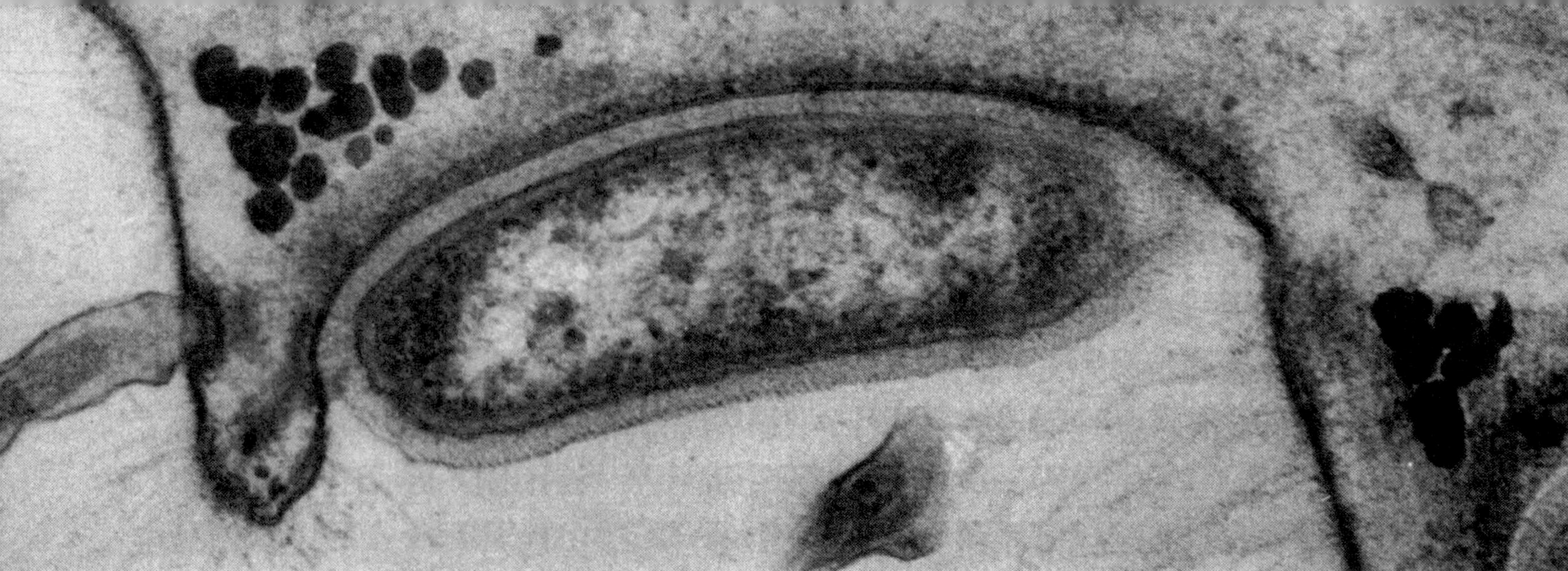

Symbiosis and the Holobiont: The Message of *Mixotricha paradoxa*

Bruce Clarke

Fig. 29 *M. paradoxa* cortex, relationships between spirochetes, cytoplasm, and bacteria. From Lynn Margulis, *Origin of Eukaryotic Cells* (New Haven: Yale University Press, 1970). © Albert V. Grimstone

Introducing the Holobiont

The late American evolutionary thinker Lynn Margulis deserves major credit for the ascending profile of the concept of the *holobiont*. Understood in this new light, every living system is embedded in a multi-organismal matrix constituted by an obligatory repertoire of reciprocal symbioses with other forms of life. The holobiont concept draws attention to the range of microbial life—bacterial, protistan, fungal—that combines in the making and maintenance of plants and animals. Margulis did not herself coin the technical term, but in her writings detailing the major evolutionary and ecological roles played by symbiotic processes, she effectively brought it to a new prominence.[1]

The concept of the holobiont challenges orthodox notions of separation among biological "individuals." This issue crosses between biological and philosophical ideas about corporeal identity.[2] The vital symbiotic dynamics brought into sharp focus by the holobiont concept were previously pushed to the margins of biology, for one reason, because they do not directly partake of traditional Darwinian matters such as competitive survival through reproductive fitness, conceptions drawing on obsolete notions of genetic individuality borrowed from non-scientific traditions. However, in the last few decades, the holobiont concept has been vetted as an authoritative scientific description of the actual ecological complexity of all eukaryotic organisms. Plant and animal holobionts

1. For the prehistory of the concept, see Jan Baedke, Alejandro Fábregas-Tejeda, and Abigail Nieves Delgado, "The Holobiont Concept before Margulis," *Journal of Experimental Zoology Part B: Molecular and Developmental Evolution* 334 (2020): 149–55, https://doi.org/10.1002/jez.b.22931. Margulis had introduced the holobiont concept into her scientific work by the early 1990s. See Lynn Margulis, *Symbiosis in Cell Evolution: Life and its Environment on the Early Earth*, 2nd ed. (San Francisco: W. H. Freeman, 1993); see also "Enter the Holobiont" in Bruce Clarke, *Gaian Systems: Lynn Margulis, Neocybernetics, and the End of the Anthropocene* (Minneapolis: University of Minnesota Press, 2020), 235–39.

2. The classic paper in this regard is Scott F. Gilbert, Jan Sapp, and Alfred I. Tauber, "A Symbiotic View of Life: We Have Never Been Individuals," *The Quarterly Review of Biology* 87, no. 4 (December 2012): 325–41, https://doi.org/10.1086/668166; see also Derek Skillings, "Holobionts and the Ecology of Organisms: Multi-Species Communities or Integrated Individuals?" *Biology and Philosophy* 31 (2016): 875–92, https://doi.org/10.1007/s10539-016-9544-0; and Javier Suárez and Vanessa Triviño, "A Metaphysical Approach to Holobiont Individuality: Holobionts as Emergent Individuals," *Quaderns de Filosofia* 6, no. 1 (2019): 59–76, https://doi.org/10.7203/qfia. 6.1.14825.

are typically analyzed into their macroscopic *hosts* and their microbial *symbionts,* although as feminist theorist Donna J. Haraway reminds us, the host is as much a symbiont of its microbes as they are of it.[3]

Writing with her eldest son, Dorion Sagan, Margulis would go on to popularize the eukaryotic microbe *Mixotricha paradoxa* as a holobiont *par excellence.* Characterizing it colorfully as the "the beast with five genomes" and the "'poster protist' for symbiogenesis," their formulations aim at evoking the genomic complexity always at hand no matter how singular an organism may appear to traditional view.[4] Symbiogenesis concerns speciation—the natural production of novel organisms—not through the happenstance viability of "random mutations" but through the viable integration of previously evolved, so to speak, pretested precursor life forms. And the multiplicity of genomes involved in the maintenance of holobiont teams bespeaks our current understanding of the *microbiome* as an ecological foundation for mutualistic symbioses of organisms across the phylogenetic spectrum. Strictly speaking, every holobiont is a community of organisms, an entire integrated ecology, a nexus of nested and interdependent beings. *Mixotricha* helps us to see that, in the biosphere altogether, complex life persists through symbioses all the way up to Gaia as a planetary system and back down to "microbial dark matter."[5] In the discussion that follows, I will detail the emergence of the protistan holobiont *Mixotricha paradoxa* as a charismatic organism in the discourse of symbiosis.

3. Donna J. Haraway, *Staying with the Trouble: Making Kin in the Chthulucene* (Durham: Duke University Press, 2016), 60.

4. Lynn Margulis and Dorion Sagan, "All for One," in Margulis and Sagan, *Dazzle Gradually: Reflections on the Nature of Nature* (White River Junction, VT: Chelsea Green, 2007), 45; originally published in *Natural History* as "The Beast with Five Genomes."

5. See Scott F. Gilbert, "Metaphors for a New Body Politic: Gaia as Holobiont," in *A Book of the Body Politic: Connecting Biology, Politics and Social Theory,* ed. Bruno Latour, Simon Schaffer, and Pasquale Gagliardi (Venice: Fondazione Giorgio Cini, 2020), 75–88; and John F. Stolz, "Gaia and her Microbiome," *FEMS Microbiology Ecology* 93, no. 2 (2017), https://doi.org/10.1093/femsec/fiw247.

Symbiotic Speculations

In 1957, at age nineteen and holding a baccalaureate from the University of Chicago, Lynn P. Alexander married Carl Sagan. They moved to Wisconsin, where he continued his graduate studies at Yerkes Observatory. She earned an MA from the University of Wisconsin in zoology and genetics in 1960, then did doctoral work in genetics between 1960 and 1963 at the University of California, Berkeley. She divorced Sagan in 1964 and received her PhD from Berkeley in 1965. From 1963 to 1967, she worked as a research assistant, lecturer, and staff member at Brandeis University in Waltham, Massachusetts. In 1966, Boston University hired her as adjunct assistant professor in the Department of Biology; in 1967, she joined their tenure track and married crystallographer Nick Margulis.

Earlier that year, after multiple rejections, she had placed a landmark article, "On the Origin of Mitosing Cells," in the *Journal of Theoretical Biology*, writing as Lynn Sagan.[6] In 1970 she published a substantial book-length version of that article's thesis, *Origin of Eukaryotic Cells*, with Yale University Press.[7] This work presented in great detail what others soon came to call "serial endosymbiosis theory," or SET, an innovatively refined yet still hypothetical account of microbial symbiogenesis, the sequential evolutionary assembly of the eukaryotic or nucleated cell from accumulated mergers among prokaryotic precursors. A few years later, Margulis participated in an episode of the BBC documentary series *Horizon* on the topic of symbiosis.[8] Aired on January 12, 1976, "Intimate Strangers" drew from popular science writings by her professional colleague, medical researcher Dr. Lewis Thomas. His celebrated 1974 work *The Lives of a Cell: Notes of a Biology Watcher* disseminated

6. Lynn Sagan, "On the Origin of Mitosing Cells," *Journal of Theoretical Biology* 14, no. 3 (March 1967): 225–74.

7. Lynn Margulis, *Origin of Eukaryotic Cells: Evidence and Research Implications for a Theory of the Origin and Evolution of Microbial, Plant, and Animal Cells on the Precambrian Earth* (New Haven: Yale University Press, 1970).

8. An excerpt from "Intimate Strangers" focused on Margulis and her spirochete theories may be viewed on YouTube: BBC, "Excerpt from BBC Horizon 'Symbiosis: Intimate Strangers,'" YouTube video, 4:58. September 29, 2017, www.youtube.com/watch?v=GwXSIlENClY.

a broadly accessible version of SET discreetly drawn from her *Origin of Eukaryotic Cells.*[9]

In an outdoor scene from "Intimate Strangers," Margulis expounds important facets of SET, in particular, the endosymbiotic origin of the mitochondrion, the eukaryotic organelle that handles the metabolism of oxygen for the cell as a whole. She visits a local pond and notes a scum of "blue-green algae" on the rocks just below the surface, a patina of prokaryotic microbes she would in other contexts be adamant about naming within its proper biological kingdom as *cyanobacteria*. These green beings are the bacterial precursors, not of mitochondria, but of chloroplasts. They do not process oxygen, but rather release it as the waste product of their evolutionary innovation, aqueous photosynthesis. In the scene at hand, they are emitting bubbles of oxygen just as their Archaean ancestors did around 2 billion years ago, when they pumped up oxygen, transforming it from a trace element to a major component of the Earth's atmosphere. While this "Great Oxidation Event" poisoned much of the biota of that era, it also set the stage for the evolution of aerobic liýe, including the bacterial precursor of the mitochondrion, a microbe that came to use the newly ambient oxygen as a radically more efficient source of cellular energy.

In this pondside scene from "Intimate Strangers," Margulis explains her scientific work for a popular audience:

> The blue-green algae are still here. Even though now they're relegated to a position of relative inconspicuousness, in fact you can just see them as a kind of scum on the rocks there. They're still around. If you look carefully down there, you can see that there are bubbles coming up out of the water. Well, the bubbles are just the oxygen, and that's just the oxygen that was responsible—that is, oxygen given off by blue-green algae was the oxygen that was responsible for the transition of the entire atmosphere. The atmosphere went from something very anaerobic, no oxygen at all, to the present level of about 20% oxygen, something like 2,000 million years ago, plus or

9. Lewis Thomas, *The Lives of a Cell: Notes of a Biology Watcher* (New York: Bantam, [1974] 1975).

> minus 100 million years. And it's just those blue-green algae—that is, their ancestors, of course—that led to that transition.
>
> So, one can imagine that early scene, where blue-green algae were giving off this enormous pollutant, that is, oxygen, into the atmosphere and poisoning just about everything around them. One of the ways of coping with the increased oxygen back then, and a way I think that was very important in the origin of the complex cells, was by symbiosis. Now, what do we mean? Two bacteria, very different kinds of bacteria, probably got together and ended up using the oxygen. The first of these bacteria simply broke down sugars and in fact couldn't use the oxygen at all. The second took the breakdown products and, with oxygen, burnt them, essentially, and derived a lot of energy from that. And the two together made a complex. And this complex, after a long period of time and many, many changes, became a very successful kind of organism. And in fact, it's that kind of organism that is the ancestor to all the higher cells. The things that were once bacteria after time became the mitochondria that we find in all of our cells—all the plant cells, all the animal cells, and even all the fungal cells.

Here Margulis asserts her once-controversial thesis that the origins of the "complex" or eukaryotic cell came about by a symbiosis between several different kinds of bacteria—in contemporary parlance, an anaerobic *Archaean* host that accepted the permanent invasion of an aerobic *eubacterium*, a prokaryote that had evolved the ability to tolerate and derive energy from oxygen. As life continued to diversify, in a further set of endosymbiotic events, photosynthetic organisms would eventually incorporate the sugar-secreting, oxygen-emitting cyanobacterium, which then evolved into another kind of eukaryotic organelle, the chloroplast. These theories challenged a prior orthodoxy regarding such organelles as differentiated offshoots of the eukaryotic nucleus. However, in the 1980s, new techniques of genetic sequencing would provide evidence for the greater part of SET, experimentally confirming the genetic markers that link both the mitochondria and chloroplasts of eukaryotic organisms back to their bacterial origins.

In 1988, Margulis went on to publish a related theoretical article in the edited collection *Speculations: The Reality Club*.[10] This invited venue gave her an occasion precisely to speculate on the ways in which her scientific interests touched upon the philosophical topic that editor John Brockman stipulated as the focus for this popular volume—the human mind. Her contribution was a remarkable discourse titled "Speculation on Speculation." The title's self-referential twist was meant quite literally: Margulis set about to speculate on the evolution of the ability to speculate: "Here I speculate on the spirochete origin of our sensory-nervous systems. A strange idea, it is easy to resist because it seems so bizarre. But because life's biochemistry and genetics are so conservative it is, I suspect, correct."[11] life's biochemistry is "conservative," that is, in the sense that whenever a metabolic capacity conferring selective advantage arises, it is seldom lost altogether.

At face value, her idea does seem peculiar. At the evolutionary base of the highest capacities of the human mind lie ... the corkscrew-shaped bacteria called spirochetes? However, these primal organisms, she explained, are some of the prime movers of microbial motility, and it is this vestigial dynamism that she saw deeply embedded in the evolution of neurons. The spirochete is distinguished for incorporating the flagellum—the whip-shaped bacterial propeller that usually protrudes outward from the cell body—on the inside, between its outer and inner cell membranes. In this strain of bacteria, multiples of such "periplasmic" flagella rotate along their length to produce powerful gyrations.[12] Margulis now asserted a direct link between their primal wriggling and mental

10. John Brockman, ed., *Speculations: The Reality Club* (New York: Prentice Hall, 1990).

11. Lynn Margulis, "Speculation on Speculation," in *Speculations*, ed. Brockman, 157–67; reprinted in Lynn Margulis and Dorion Sagan, *Dazzle Gradually: Reflections on the Nature of Nature* (White River Junction, VT: Chelsea Green, 2007), 48; citations are from the 2007 version.

12. See Lynn Margulis, "Spirochetes Awake: Syphilis and Nietzsche's Mad Genius," in Margulis and Sagan, *Dazzle Gradually*, 59–60.

operation: "Movement itself is an ancestral bacterial trait, and thought, I am suggesting, is a kind of cell movement."[13]

Throughout the 1980s, Margulis enjoyed a series of contacts with the neurobiological theorist Francisco J. Varela, coinventor with Humberto Maturana of the concept of autopoiesis and, at that time, the lead author of a 1991 volume titled *The Embodied Mind*.[14] Margulis's spirochetal speculations radically extended a similar register of embodied cognition. But at root, she built her theories about the "spirochete origin" of the functionality of the neuron upon the same chain of evolutionary hypotheses guiding her signature work on the evolution of the "complex" cell. Here spirochetes also played a pivotal but different role, in this context as endosymbionts within the primal eukaryotic consortium. Some of the tiniest creatures of the biosphere, Margulis now theorized, also had an abiding hand in the emergence and expansion of human brains, whose lofty minds are capable of grasping at the heavens. Here is Margulis in "Speculation on Speculation," contemplating this newly concrete, literally embodied version of the correspondence between the microcosm and the macrocosm:

> My speculations, two thousand million years later, may be the creative outcome of an ancient uneasy peace. If this reckoning is true, then the spirochetal remnants may be struggling to exist in our brains, attempting to swim, grow, feed, connect with their fellows, and reproduce. The interactions between these subvisible actors, now full member-components of our nerve cells, are sensitive to the experience we bring them. Perception, thought, speculation, and memory, of course, are all active processes; I speculate that these are the large-scale manifestations of the small-scale community ecology, that is, the fusion of two ancient forms of bacteria.[15]

13. Margulis, "Speculation on Speculation," 49.

14. For details, see Clarke, *Gaian Systems*, 143–56; Francisco J. Varela, Evan Thompson, and Eleanor Rosch, *The Embodied Mind: Cognitive Science and Human Experience* (Cambridge: MIT Press, 1991).

15. Margulis, "Speculation on Speculation," 52.

"An Earnest Proposal"

Following the path of the spirochete brings us directly to Lewis Thomas's contemporaneous appreciation of Margulis's original orchestration of serial endosymbiosis theory in *The Lives of a Cell.* This work collected Thomas's monthly columns published between 1971 and 1973 in the *New England Journal of Medicine*. Of the many chapters in this slim volume, I will focus here solely on one, "An Earnest Proposal."[16] Its title alludes to Jonathan Swift's 1729 broadside, "A Modest Proposal for Preventing the Children of Poor People From Being a Burthen to Their Parents or Country, and for Making Them Beneficial to the Publick." Herein, Swift famously channeled his outrage at the poverty of the Irish under British rule into a mock proposal that the Catholic poor should support themselves by selling their superfluous babies for stew meat and fine leather. Thomas's "Earnest Proposal" imagines another kind of holocaust, coming forward not in tones of savage satire but in a grim mode of mordant irony, alleviated by an inspired set of offbeat scientific details.

Originally published in 1971, "An Earnest Proposal" begins by noting that, from dating services to the control of nuclear arsenals, the affairs of modern life are being increasingly supervised by computers. Thomas muses sardonically that, before too long, "various networks will begin to touch, fuse, and then, in their coalescence, they will start sorting and retrieving each other, and we will all becomes bits of information on an enormous grid."[17] Beyond his prescient view of globalizing networks, however, he considers the most concerning problem to be the programmatic computer misprogramming of basic assumptions about reality. If the computers apparently taking charge of our world "are programmed to regulate human behavior *according to today's view of nature*, we are surely in for apocalypse."[18] Thomas asserts that the popular contemporary view of life is based on a defective construction of the theory of

16. Lewis Thomas, "An Earnest Proposal," in *The Lives of a Cell*, 29–34; originally published in *New England Journal of Medicine* 285, no. 20 (November 11, 1971): 1132–33.

17. Ibid., 29.

18. Ibid., my italics.

evolution. In the midst of a hot war in Vietnam and an ongoing Cold War between the superpowers, Thomas expresses dismay at the prospect of social Darwinist notions fueling the nightmare of an all-out nuclear exchange:

> The men who run the affairs of nations today are, by and large, our practical men. They have been taught that the world is an arrangement of adversary systems, that force is what counts, aggression is what drives us at the core, only the fittest can survive, and only might can make more might. Thus, it is in observance of nature's law that we have planted, like perennial tubers, the numberless nameless missiles in the soil of Russia and China and our Midwestern farmlands, with more to come, poised to fly out at a nanosecond's notice, and meticulously engineered to ignite, in the centers of all our cities, artificial suns. If we let fly enough of them at once, we can even burn out the one-celled green creatures in the sea, and thus turn off the oxygen.[19]

Ironically echoing Swift's "modest" plea for social justice in a British colony, Thomas's satire conveys a reasonable malaise over anthropogenic threats to the very survival of human life, or even of aerobic life altogether. In this specific note of a "biology watcher," Thomas looks out at the military-technological landscape of the Nixon-Brezhnev era and professes to detect a hypertrophy of abhorrent evolutionist dogma, the bogus "nature" of a flawed scientific ideology, being drummed into the minds of "practical men" and, through them, programmed into the mindless biases of computational databases.

This setup for Thomas's earnest proposal conveys broad misgivings over a world darkened by nuclear armaments. He traces their proliferation to the skewing of human affairs by a falsely predatory view of the nature of life. Nor will our deference to impersonal cybernetic controls save us: our computers simply reproduce and compound our own unsound convictions. So then,

19. Ibid., 30.

Thomas proposes that, before misplaced beliefs in humanity's natural depravity become self-fulfilling prophecies of nuclear annihilation, and in view of the information banks that human and artificial intelligences now draw upon to make decisions about life and death, we should try to correct them in at least one key instance. Having raised this not entirely mock alarm, he offers a strategy to buy the world some time before the decision-makers do something rash and trigger some irrevocable event of unfathomable destruction: "I suggest that we defer further action until we have acquired a really complete set of information concerning at least one living thing. Then, at least, we shall be able to claim that we know what we are doing."[20] Here is the core of his "earnest proposal": let us pause hostilities long enough to take the time needed to study comprehensively one life form that represents the truth of nature at large. But what creature could be so emblematic of biological universality that, once our comprehension of its way of being has been grasped by every informed mind and uploaded into every mainframe, a conflicted humanity would then have a fighting chance to compute its own coexistence? Let it be—*Mixotricha paradoxa*. But of all the exemplary animals and admirable plants, of all the versatile fungi and hardy lichens on Earth, why does Thomas single out a fabulously obscure amitochondriate protist, a polymastigote flagellate covered with varieties of spirochetes and found only in the hindgut of a single species of Australian termite, *Mastotermes darwiniensis*?

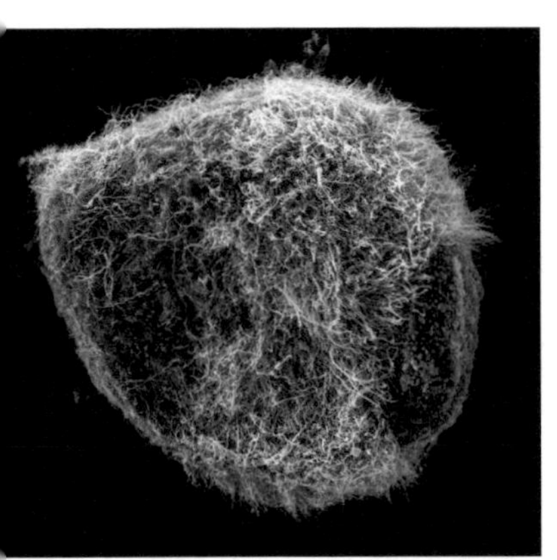

Fig. 30 M. paradoxa, date unknown. © Lynn Margulis Laboratory

The Message of *Mixotricha*

Let us trace this thread back to its origins. As reported in her 1933 article "Protozoa from Australian Termites," microbiologist Jean L. Sutherland first discovered *Mixotricha paradoxa* and named it as such due to its "paradoxically

20. Ibid.

mixed-up hairs."[21] The salient, largely descriptive details Sutherland gives about this "large flagellate parasite" concern its hirsute oddities and the "presence of cilia as locomotor organs":

> The body is invested with a coat of cilia disposed in closely packed transverse bands. Their insertion and movement are quite different from those of the short flagella of Trichonymphids, and they can definitely be distinguished from the foreign organisms which invest such forms as Pseudodevescovina by the nature of their movement, and by their basal granules situated in the cortical zone … . In general appearance, and particularly in the presence of both cilia and flagella, this form is quite unlike any described termite parasite.[22]

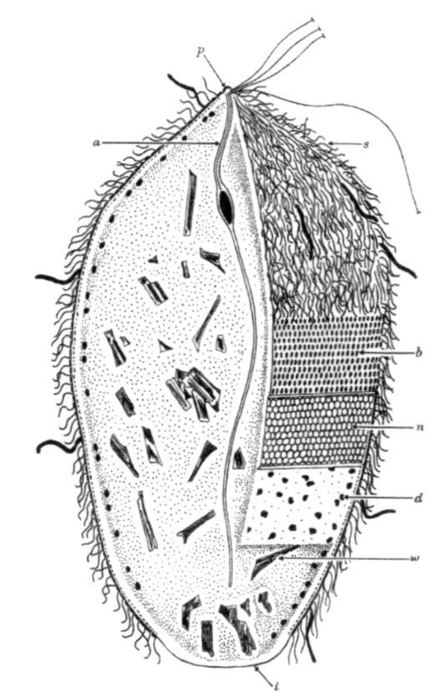

Fig. 31 Diagram of *M. paradoxa*, 1964. © Lemuel R. Cleveland and Albert V. Grimstone

Although it is typically *flagella* that move microbes around, Sutherland concludes that the "locomotor organs" that coat *Mixotricha* are, most unusually, "true cilia," that is, not parasitic "foreign organisms" but organelles expressed by the protist's own genome. This identification is driven by her assumption that uniform movements sufficiently well-coordinated to provide effective motility can arise only from the "true" organs of a unitary organism. However, putting that curiosity aside, Sutherland regards the newfound *Mixotricha* itself to be a mere internal parasite. One is reminded of Jonathan Swift's "On Poetry: A Rhapsody" of 1733: "So nat'ralists observe, a flea / Has smaller fleas that on him prey; / And these have smaller fleas to bite 'em. / And so proceeds *Ad infinitum*." In her rendering, *Mixotricha* is simply a kind of internal flea that happens to have some noteworthy quirks. Otherwise, it just floats about freeloading on the cellulose in the termite's hindgut. Following the science of her moment, she categorizes *Mixotricha* as

21. Jean L. Sutherland, "Protozoa from Australian Termites," *Quarterly Journal of Microscopical Science* 76 (1933): 145–73.

22. Ibid., 163, 165.

a parasitic opportunist rendering nothing back to its host. It is not *this* Mixotricha on which Lewis Thomas would hang his hopes for mankind.

Three decades later, microscopists Lemuel R. Cleveland and Albert V. Grimstone published "The Fine Structure of the Flagellate *Mixotricha paradoxa* and its Associated Micro-Organisms."[23] In this article, they note that the study of this organism has lain dormant since Sutherland's paper, despite its possession of "a feature without parallel among the Protozoa, namely the presence of both a number of long anterior flagella and a complete covering of short cilia," amounting to an "apparent occurrence of two kinds of locomotory organelle."[24] However, Cleveland and Grimstone conclude, Sutherland erred in her description of the "paradoxically mixed-up hairs" of *Mixotricha paradoxa*. They now come at their subject with electron microscopy, enabling high-resolution imagery at much finer scales than those possible with Sutherland's light microscope: "it is clear that these structures are not cilia but adherent spirochetes. They do not move in the manner of cilia, with alternate effective and recovery strokes, but instead undulate, exactly as spirochetes do … . It was the apparent presence of basal granules which led her to identify these structures as cilia rather than spirochetes … . Evidence as to the nature of the granules cannot readily be obtained by light microscopy, but it will be shown below that they are adherent bacteria."[25]

Cleveland and Grimstone now observe *Mixotricha* more closely as a consortium in which a protistan eukaryote is all mixed up with radically different strains of bacteria. The earlier misrecognition of its composition was due in part to this unforeseen composite of eukaryotic and prokaryotic participants. Observation at finer resolution has eliminated the earlier "paradox"—the seemingly

23. Lemuel R. Cleveland and Albert V. Grimstone, "The Fine Structure of the Flagellate *Mixotricha paradoxa* and its Associated Micro-Organisms," *Proceedings of the Royal Society of London, Series B, Biological Sciences* 159, no. 977 (March 17, 1964): 668–86.

24. Ibid., 668.

25. Ibid., 670–71.

redundant co-presence of two distinct kinds of locomotory organelles, both cilia and flagella. However, now another conundrum emerges: attached to the protist's cortex by association with yet another kind of adherent bacterium, a profuse complement of individual spirochetes somehow provides effective locomotory services. But how could these prokaryotic hitchhikers achieve such startlingly effective coordination despite having no phenotypic connection to their eukaryotic host? Yet there it is, observable in real time: *Mixotricha's* spirochetal followers have set up a cross-kingdom and cross-domain regime of cooperative association, indeed, of effective coordination without a centralized organ conducting the orchestration. Cleveland and Grimstone can only speculate that the aggregation of spirochetes itself may constitute a "pacemaker ... by virtue of their dense packing ... near the papilla, where there may be as many as four per bracket, ... able to impose their timing on those located more posteriorly."[26] With Cleveland and Grimstone's hybridic and collaborative *Mixotricha*, we have moved a few steps closer to Thomas's exemplary organism. Moreover, *Mixotricha*'s adherent spirochetes—a synchronized bacterial throng, much like a school of fish or a flock of birds—appear to exhibit collective behavior. Is this a microbial case of distributed mind, or at least, of bacterial sensitivity to their environment sufficient to achieve effective sensory-motor inter-coordination?

Be that as it may, neither Sutherland's nor Cleveland and Grimstone's treatments of *Mixotricha* make any mention of the concept of symbiosis. In historical context, this absence is not curious; rather, it may be taken as marking the marginal status of that concept as late as in the 1960s.[27] In contrast, due to her pioneering renovation and development of that topic at that same moment, *Origin of Eukaryotic Cells* treats *Mixotricha* as an occasion for an emphatic discussion of symbiosis and symbiogenesis. Indeed, Margulis's 1970 text is the sole source Thomas himself cites in "An Earnest Proposal" for his knowledge regarding *Mixotricha* and its exotic peculiarities. The epigraph to Chapter 6 of *Origin of Eukaryotic Cells*, titled

26. Ibid., 682.

27. See Jan Sapp, *Evolution by Association: A History of Symbiosis* (Oxford: Oxford University Press, 1994).

"Symbiosis," is a long quotation from Cleveland and Grimstone's 1964 paper, focused specifically on the superior motility that arises from *Mixotricha*'s complement of spirochetes. "Except when obstructed," Cleveland and Grimstone write:

> *Mixotricha* glides along uninterruptedly, at constant speed and usually in a straight line. These characteristics ... are markedly different from those of other large flagellates .... *Mixotricha* is propelled not by its flagella but by the undulations of its attached spirochaetes. The chief evidence for this assertion is that in all actively swimming individuals the spirochaetes undulate vigorously and are well co-ordinated .... There is, in general, a clear correlation between the degree of activity and coordination of the spirochaetes and the speed of movement of their host. In spite of the closest study no other possible means of locomotion has been detected.[28]

It was Margulis who would then proceed to bring in the symbiotic connections we now take to be obvious considerations.

In our present moment, the broad dissemination of knowledge about symbiosis and the general acceptance of its vital role in the biosphere owes something to Thomas's appreciative popular treatments and a great deal more to Margulis's daring scientific innovations. When described in this current biological dialect, the spirochetes that adhere to *Mixotricha* are termed *ectosymbionts*. The functional outcome of their participation with their protistan host is a *motility symbiosis*. Thus, in its full ecological complexity, *Mixotricha* presents itself as a holobiont hosting mutualistic relations with its spirochetal and other bacterial ecto- and endosymbionts. Its termite host in its turn is thus a higher-order holobiont in a mutualistic relationship with *Mixotricha* as an *endosymbiont*. *Mixotricha* breaks down the cellulose that the termite ingests into nutrients that the termite can actually digest. In return, it takes its own share of the hindgut table and passes on the leavings to its drivers, the spirochetes.

28. Cleveland and Grimstone, cited in Margulis, *Origin of Eukaryotic Cells*, 143.

Margulis's own discussion of *Mixotricha* in *Origin of Eukaryotic Cells* begins with that striking epigraph marshaling *Mixotricha* as a delegate for symbiosis in general. Her text renders this complex protist especially memorable with a striking selection of Cleveland and Grimstone's spectacular electron micrographs, showing various sections of *Mixotricha*'s ridged cortex, where two different eubacterial ectosymbionts come together in regular tiers at complex attachment sites. We see that the protist's own body has evolved so as to provide docking locations for its bacterial confederates. We have now fully arrived at the exemplary organism of Thomas's "Earnest Proposal." The ecological arrangements of this microcosmic consortium assemble a geopolitical allegory in which diverse participants play their part, hold up their end, and receive a living share. Everybody eats, and life is good.

As I examine Margulis's stunningly detailed and daring 1970 text through the lens of Thomas's brief popular appreciation, I see him using her foregrounding of *Mixotricha* to highlight key aspects of her book's larger arguments. Moreover, he extends those arguments in a way that renders their wider implications compelling and unforgettable. I speculate that it may have been Thomas's attentions here that first confirmed for Margulis how *Mixotricha paradoxa* could become universally emblematic. At the same time, thanks to the sheer heuristic potency of the evolutionary theories Margulis had already put on the table, together with his hunch that she was mostly right, Thomas was able to prepare this nugget of bioscience for the rhetorical feast that is "An Earnest Proposal."

As we noted before, *Origin of Eukaryotic Cells* contested an earlier evolutionary consensus. This prior construction held that the eukaryotic cell arose due to an accumulation of random single-point mutations selected for in some prokaryote, somehow giving rise to a nucleus binding the cell's genome, and that the other eukaryotic organelles then emerged by gradual differentiation from that nucleus. Margulis countered with a symbiogenetic explanation, viewing the evolution of the eukaryotic cell as the result of the multiple assembly and functional aggregation of different free-living prokaryotic precursors. Thanks to genome sequencing

Fig. 32 *M. paradoxa* cortex, bacteria and symbiotic spirochetes. From Lynn Margulis, *Origin of Eukaryotic Cells* (New Haven: Yale University Press, 1970). © Albert V. Grimstone

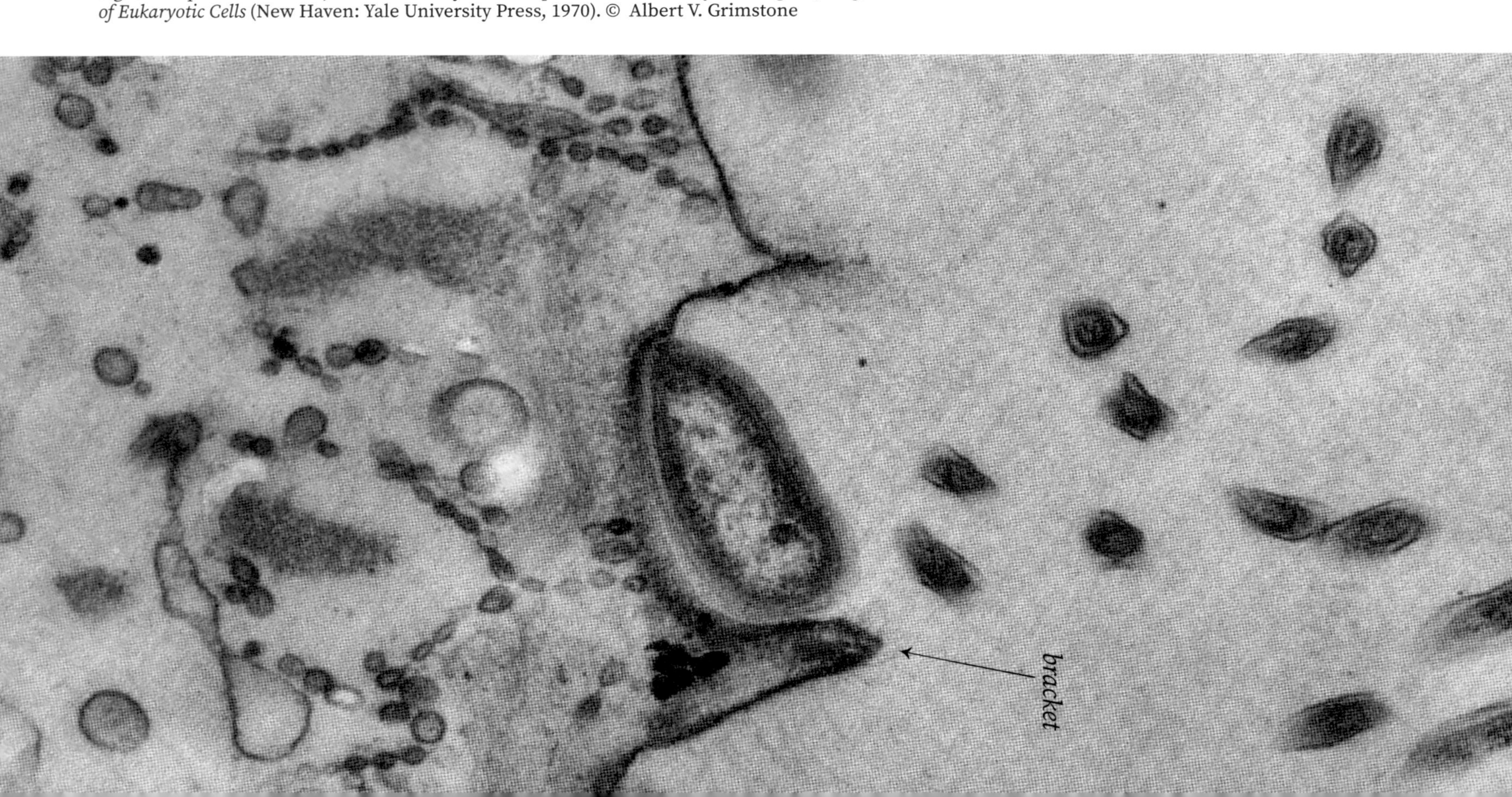

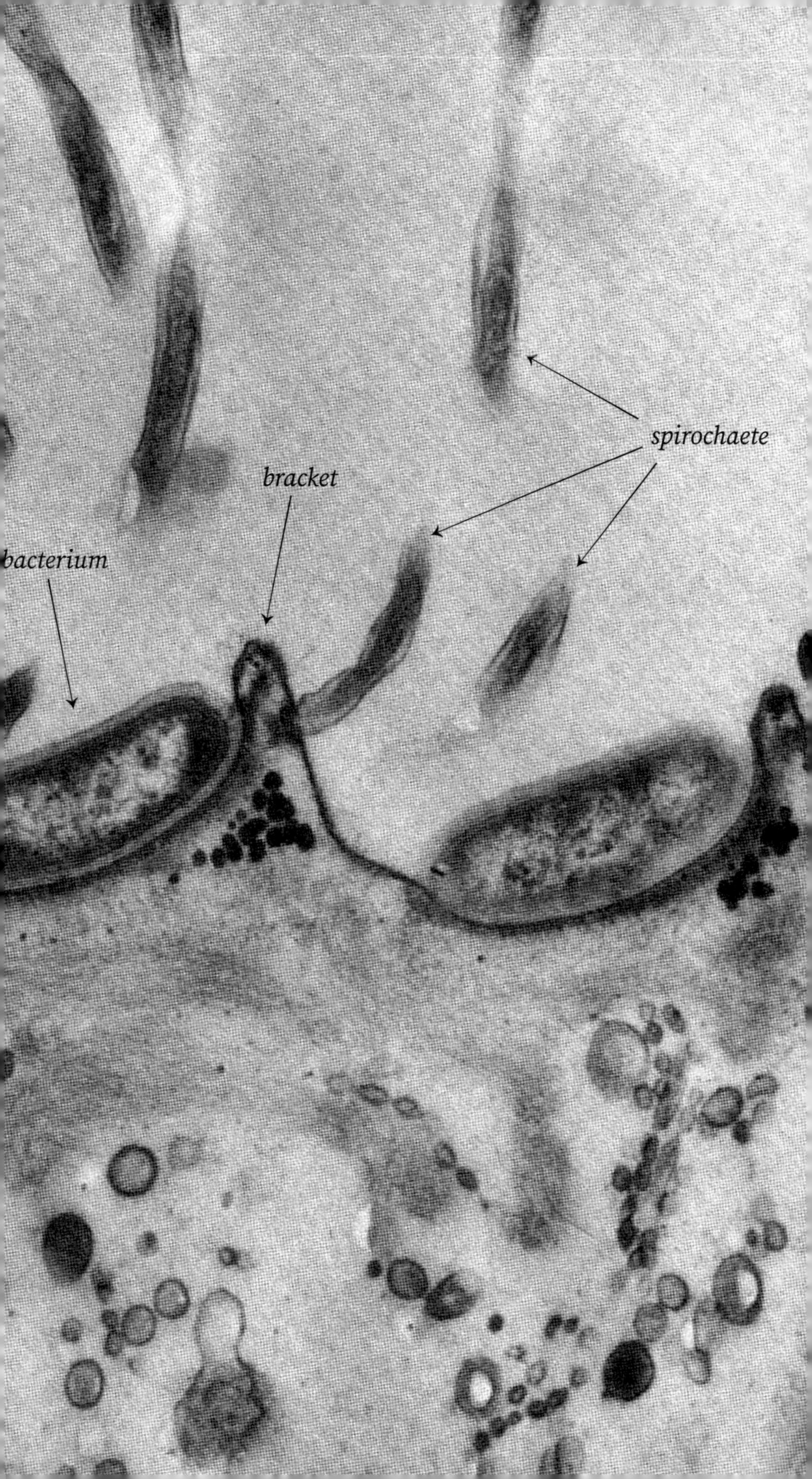
spirochaete
bracket
bacterium

methods that arrived only a decade or so after Margulis's comprehensive theorizations at the turn of the 1970s, it is now textbook science that the eukaryotic mitochondrion and chloroplast originated in this way. Their ancestors entered the evolving eukaryotic consortium intact and then, one by one, transitioned from endosymbionts into organelles by developing obligate genetic and metabolic interrelations with the cell as a complex system. In addition, she theorized, a pre-mitotic ur-eukaryote eventually evolved the reproductive machinery of mitosis. This event would be yet another adaptive outcome of the symbiogenetic acquisition of spirochetal motility turned inward to conduct the moving parts of the mitotic process. *Origin of Eukaryotic Cells* develops this intricate theory in book length.

Although Thomas's avid transmission of Margulis's suite of innovations in that text is scattered throughout *The Lives of a Cell*, "An Earnest Proposal" provides a synopsis of her total theory, and by an act of creative reception, focuses that theory entirely upon *Mixotricha*. Moreover, Thomas adds ecological details that both extend the symbiotic theme beyond Margulis's own evolutionary arguments at that time and supply key functional details missing from the accounts of the earlier investigators. He emphatically underscores the mutualistic symbiosis shared between the termite host and its so-called parasites: "In the termite ecosystem, an arrangement of Byzantine complexity," he writes, *Mixotricha* "stands at the epicenter. Without him, the wood, however finely chewed, would never get digested; he supplies the enzymes that break down cellulose to edible carbohydrate."[29] Thomas then notes matters that we have already reviewed, such as how presumed locomotory organelles viewed at the requisite scale are in fact shown to be distinct but adherent spirochetes, "outsiders, in to help with the business," alongside other ecto- and endosymbiotic bacteria, "living in symbiosis with the spirochetes and the protozoan."[30] Then in one clean and lucid stoke, Thomas cuts through to the unique glamor of *Mixotricha*: "The whole animal, or ecosystem, *stuck for*

29. Thomas, "An Earnest Proposal," 31.

30. Ibid., 32.

the time being halfway along in evolution, appears to be a model for the development of cells like our own."[31]

Rescued by intrepid microscopists and phylogenetic systematists from base obscurity, unkempt and wearing its abject hybridity and symbiogenesis-in-progress on its sleeve, and contrary to an exclusively competitive-predatory view of living relations, what Margulis implies and Thomas makes entirely explicit is that *Mixotricha* models nature's good nature: "If it is in the nature of living things to pool resources, to fuse when possible, we would have a new way of accounting for the progressive enrichment and complexity of form in living things."[32] Margulis spent the rest of her working life discerning the tracks of the spirochete while working out further details for her larger account of symbiogenetic evolutionary innovation.[33] Scaling up what Margulis more carefully presents as "the symbiont theory," Thomas draws out the planetary significance of the microcosm. Working directly from Margulis's own speculations about the evolutionary force of symbiosis, Thomas's text, in a brilliant intuition, appears to prefigure the contemporary concept of *sympoiesis* in its Gaian dimensions:

> There is an underlying force that drives together the several creatures comprising myxotricha [*sic*], and then drives the assemblage into union with the termite. If we could understand this tendency, we would catch a glimpse of the process that brought single separate cells together for the construction of metazoans, culminating in the invention of roses, dolphins, and, of course, ourselves. It might turn out that the same

31. Ibid., my italics.

32. Ibid., 33.

33. Further research has not borne out Margulis's spirochete thesis in the form she left it. Putting the evolution of neurons aside, what she saw as the spirochete's contribution to eukaryosis was the tubulin machinery from which evolved the 9+2(2) microtubules of the kinetosomes involved in the mitotic apparatus. It now appears that the tubulin in question was donated by the archaeal host of the eubacterial consortium. See Natalya Yutin and Eugene V. Koonin, "Archaeal Origin of Tubulin," *Biology Direct* 7, no. 10 (2012): 1–9; and Anja Spang et al., "Complex Archaea that Bridge the Gap between Prokaryotes and Eukaryotes," *Nature* 521 (2015): 173–79, https://doi.org/10.1038/nature14447. I am grateful to Betsey Dexter Dyer and John Boyd for these references.

> tendency underlies the joining of organisms into communities, communities into ecosystems, and ecosystems into the biosphere.[34]

It is no small thing for Thomas's inspired take on the symbiotic status of *Mixotricha* to assemble in one passage anticipations of coevolutionary collaboration throughout the biosphere, and to do so in the midst of an independent sketch for a Gaia hypothesis that, in 1971, had not yet been publicly christened with a proper name and was largely still a matter of private correspondence between James E. Lovelock and a thirty-three-year-old professor of biology at Boston University named Lynn Margulis.[35] It is equally extraordinary that, in this same passage, still riffing on the symbiotic status of *Mixotricha*, Thomas anticipates the demise of the old self-defensive immunitary paradigm that ran in tandem with militarized versions of "the survival of the fittest." The virtually biopolitical continuation of Thomas's peroration in "An Earnest Proposal" may be favorably compared to the newer immunology spearheaded in the 1980s by Francisco J. Varela and continuing in important work detailing the extent to which animal immune systems are in fact controlled by their symbiotic microbiota.[36] "If this is, in fact, the drift of things, the way of the world"—this being the sheer normality of symbiotic and sympoietic assemblage throughout the biosphere—then "we may come to view immune reactions, genes for the chemical marking of self, and perhaps all reflexive responses of aggression and defense as secondary developments in evolution, necessary for the

34. Thomas, "An Earnest Proposal," 33; on sympoiesis, see Bruce Clarke and Scott F. Gilbert, "Margulis, Autopoiesis, and Sympoiesis," in *Symbionts: Contemporary Artists and the Biosphere*, ed. Caroline A. Jones, Natalie Bell, and Selby Nimrod (Cambridge, MA: MIT Press, 2022), 63–77; and Bruce Clarke, "Gaian Technics: Lynn Margulis, Natural Technicity, and the Technosphere," in *The Space of Technicity: Theorizing Social, Technical and Environmental Entanglements*, ed. Robert A. Gorny, Stavros Kousoulas, Dulmini Perera, and Andrej Radman (Delft: TU Delft OPEN Publishing in partnership with Jap Sam Books, 2024), 117–45.

35. See Bruce Clarke and Sébastien Dutreuil, eds., *Writing Gaia: The Scientific Correspondence of James Lovelock and Lynn Margulis* (Cambridge: Cambridge University Press, 2022).

36. See Margaret McFall-Ngai et al., "Animals in a Bacterial World, a New Imperative for the Life Sciences," *PNAS* 110, no. 9 (February 26, 2013): 3229–36; and Clarke, "Planetary Immunity," in *Gaian Systems*, esp. 224–39.

regulation and modulation of symbiosis, not designed to break into the process, only to keep it from getting out of hand."[37]

Coda: The Microcosmic Dance

Microbes are measured in microns—millionths of a meter. Larger protists—eukaryotic microbes such as *Mixotricha*—may span hundreds of microns, but most prokaryotic microbes—archaea and bacteria such as spirochetes—measure in single units. We saw that this discrepancy in spatial scale at first impeded the recognition of *Mixotricha* as a holobiont, a being in which five different microorganisms consort, four bacterial symbionts permanently nestling within or upon a protist host. The American microbiologist and science-fiction author Joan Slonczewski's 2000 novel *Brain Plague* offers a comparable literary depiction of such scalar discrepancies among disparate organismal dimensions. The human characters of that far-future storyworld cultivate a symbiosis with sentient alien prokaryotes called "micros." Once taken into the bodies of their human hosts, these "brain enhancers ... like mental mitochondria" set up tiny cities "in the arachnoid, a web of tissue between the outer linings of the brain." At one level, the story told in *Brain Plague* negotiates the differences in *temporal* perception between micros and humans: "Micros live ten thousand times faster than we do. For them, one minute feels like a week. An hour is a year; a day is a generation."[38]

Moreover, regarding the divergence in *spatial* scale among living organisms, to a symbiotic microbe, the plant or animal body of its host, even that of a termite, is an entire ecosystem—it might as well be a biosphere unto itself. In the precursor text to *Brain Plague* within Slonczewski's four-novel *Elysium Cycle*, *The Children Star* tells how the sentient micros of *Brain Plague* originally colonized the bodies of and then established abstract linguistic communication

37. Thomas, "An Earnest Proposal," 33.

38. Joan Slonczewski, *Brain Plague* (New York: Tor, 2000), 18, 32, 33; see also Bruce Clarke, "Posthuman Narration in the Elysium Cycle," in *Posthuman Biopolitics: The Science Fiction of Joan Slonczewski*, ed. Bruce Clarke (New York: Palgrave Macmillan, 2020), 17–45.

with the human colonizers of their home planet. In one passage of *The Children Star*, a particular sect of micros, "the Dancing People," having set up residence within their new human hosts, express wonderment at their good fortune. Previously, they had evolved as dwellers within one or another native life-form that they could readily submit to their internal domestication. But now, with the arrival of alien humans within their biosphere at large, they have come into possession of a host organism as intelligent as they are. Having taught their astonished first human carrier the basics of their language, they inform him how tidings came to them "about different kinds of worlds," that is, the human explorers themselves:

> The Dancing People marveled. How could such a world appear; where did they come from? ... To seek answers, some of us took to the whirrs and braved the passage into the alien worlds. Many died, for the habitat was harsh and unforgiving, its physiology foreign to our control. But we learned and adapted ... . Then we discovered an amazing thing: The alien world had intelligent feelings. You could, in fact, understand us, responding to our most intimate desires.[39]

In Slonczewski's lively scenario, intelligent microbes like extraplanetary astronauts set about to explore and exploit animal bodies arriving from another world altogether, only to discover that these new worlds have minds of their own.

The conclusion to Margulis's "Speculation on Speculation" foreshadows how these science fictions operate by means of a microbial doubling of the inquisitive activities of the human characters investigating the alien organisms they have taken possession of, but which in fact have taken possession of them. *The Children Star* and *Brain Plague* depict a mutuality of consciousness shared out between beings at the microbial and human scales. Margulis's evolutionary speculations in the immediately preceding decades propose the propriety of conceiving of ourselves and of the

39. Joan Slonczewski, *The Children Star* (New York: Tor, 1998), 282.

microbial natures of which we are composed as similarly sentient through and through:

> All I suggest is that we compare consciousness with spirochete microbial ecology. We may be vessels, large ships, unwitting sanctuaries to the thriving communities comprising us. When they are starved, cramped, or stimulated we have inchoate feelings. Perhaps we should get to know ourselves better. We might then recognize our speculations as the dance networks of ancient, restless, tiny beings that connect our parts.[40]

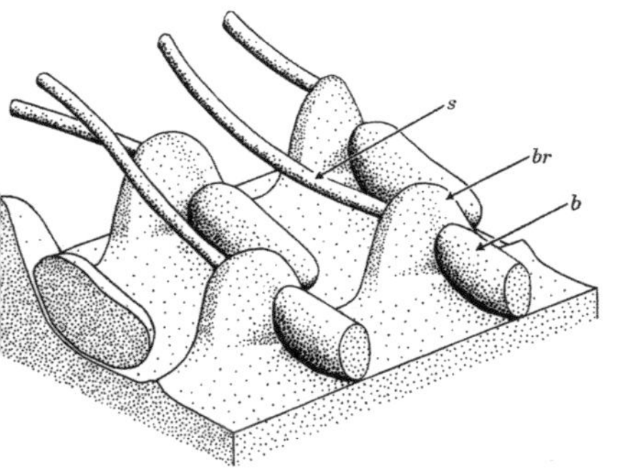

rial symbionts, 1964. © Lemuel R. Cleveland and Albert V. Grimstone

At the end of "An Earnest Proposal," Thomas imagines that, once another decade's study of *Mixotricha paradoxa* has been programmed into the master computer that also holds all the codes to the deployment of the nukes, "the feeding in of all the information then available will result, after a few seconds of whirring, in something like the following message, neatly and speedily printed out: 'Request more data. How are spirochetes attached? Do not fire.'"[41] ∞

Digital Content 04 Lynn Margulis, Bruce Clarke, Dorion Sagan, and Spherical, *untitled*, 2022.

40. Margulis, "Speculation on Speculation," 55. It is fascinating to note that her daring speculations of several decades ago have now resurfaced as a serious, if still controversial, thesis. See Arthur S. Reber, Frantisek Baluska, and William Miller, *The Sentient Cell: The Cellular Foundations of Consciousness* (Oxford: Oxford University Press, 2024).

41. Thomas, "An Earnest Proposal," 34; for more on spirochete attachment sites, see Lynn Margulis and Ricardo Guerrero, "Two Plus Three Equal One: Individuals Emerge form Bacterial Communities," in *Gaia 2—Emergence: The New Science of Becoming*, ed. William Irwin Thompson (Hudson, NY: Lindisfarne Press, 1991), 50–67.

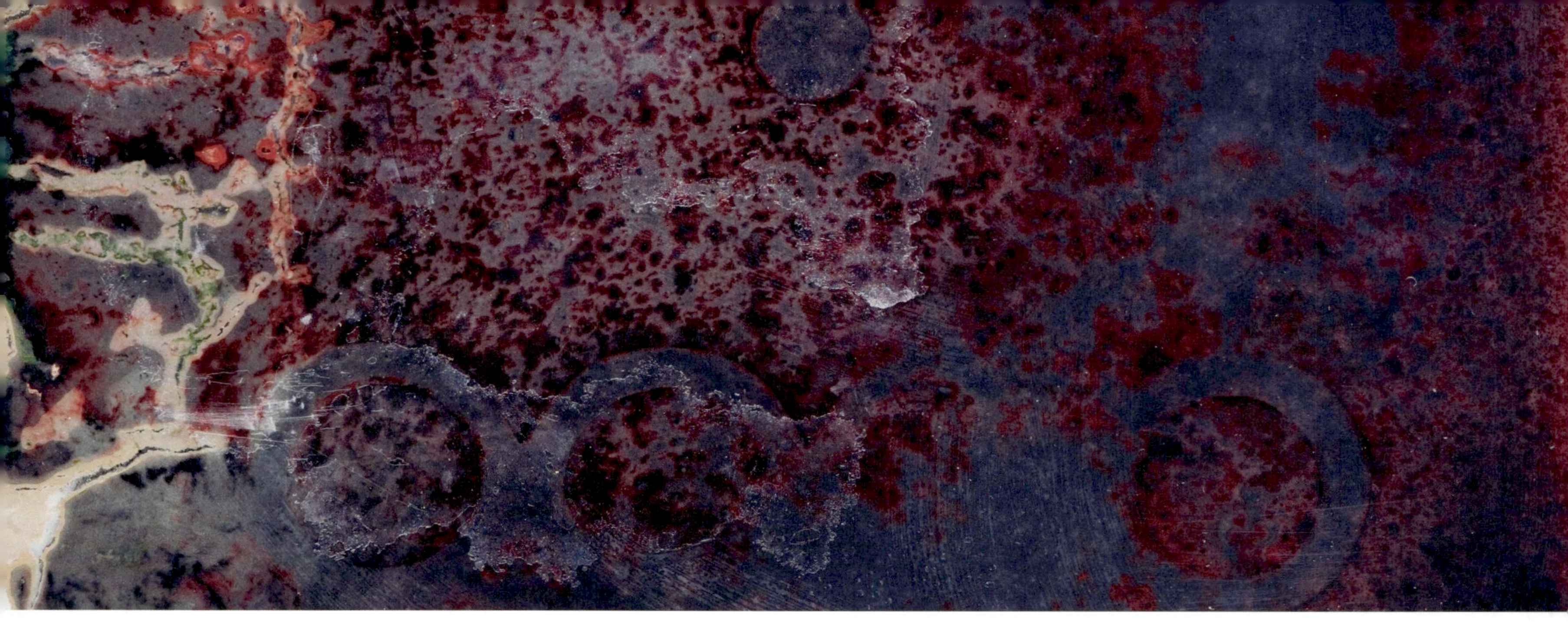

Specimen of Secrecy about Marvelous Discoveries

Eduardo Kac

Fig. 34 Eduardo Kac, *Theorem*, biotope, 2006. Private collection, Rio de Janeiro. © Eduardo Kac

Specimen of Secrecy about Marvelous Discoveries is a series of works comprising what I refer to as "biotopes," i.e., living pieces that change during the exhibition in response to internal metabolic conditions and environmental circumstances, including temperature, relative humidity, airflow, and the light levels in the exhibition space. Each of my biotopes is literally a self-sustaining ecology consisting of thousands of very small living beings in a medium of earth, water, and other materials. I orchestrate the metabolism of this diverse microbial life in order to produce my constantly evolving living works. *Specimen of Secrecy about Marvelous Discoveries* was first exhibited at the Singapore Biennale in 2006.[1]

My biotopes expand on ecological and evolutionary issues that I have previously explored in transgenic works such as *The Eighth Day* (2001). At the same time, the biotopes further develop dialogical principles that have always been central to my work.

The biotopes are a discrete ecology because, within their world, the microorganisms interact with and support each other (that is, the activities of one organism enable others to grow, and vice versa). However, they are not entirely secluded from the outside world: the aerobic organisms within the biotope absorb oxygen from the outside (while the anaerobic organisms comfortably migrate to regions that the air cannot reach). A complex set of relationships emerge as the work unfolds, bringing together internal dialogical interactions between the microorganisms in the biotope and the interactions between the biotope as a discrete unit and the external world.

The biotope is what I call a "nomadic ecology," that is, an ecological system that interacts with its surroundings as it travels around the world. Every time a biotope migrates from one location

1. Works from the *Specimen of Secrecy about Marvelous Discoveries* series were exhibited as follows: Singapore Biennale 2006 (September 4 to November 12, 2006); *Hodibis Potax,* solo show, Galerie de la Médiathèque Elsa Triolet à Villejuif, Paris (May 22 to June 1, 2007); *Du sonore et du visuel 2,* Galerie In Situ / Fabienne Leclerc, Paris (June 2 to July 31, 2007); *Fringe Exhibitions,* solo show, Los Angeles (September 8 to October 6, 2007); *Panorama 12,* Le Fresnoy, Studio National des arts contemporains, Tourcoing, France, (August 12 to 31, 2009); *Eduardo Kac: Lagoglyphs, Biotopes and Transgenic Works,* solo show curated by Christiane Paul, Oi Futuro, Rio de Janeiro, Brazil (January 25 to March 30, 2010); *Doppler Effect,* Kunsthalle zu Kiel, Kiel, Germany (January 30 to May 2, 2010).

to another, the very act of transporting it causes the unpredictable redistribution of the microorganisms inside (due to the constant physical agitation inherent in the course of a journey). Once in place, the biotope self-regulates with internal migrations, metabolic exchanges, and material settling. An extended stay in a single location might yield different behavior, possibly resulting in regions of settlement and color concentration. The biotope is affected by a number of factors, including the very presence of viewers, which can increase the temperature in the room (warm bodies) and release other microorganisms into the air (breathing, sneezing).

The exhibition opening is the birth of a given biotope. Once an exhibition begins, I allow the microorganisms in suspended animation to become active again. From that point on, I no longer intervene. The work becomes progressively different, changing every day, every week, every month.

When the viewer looks at a biotope, she sees what could be described as an "image." Indeed, it seems to comprise a whole repertoire of visual procedures, such as rips, blurs, scratches, warps, slashes, composites, scrawls, color manipulations, smears, gashes, shadows, gougings, scalings, scrapings, abrasions, transparencies, inscriptions, and overpastings. However, since this "image" is always evolving into its next transformative state, the perceived "stillness" is more a consequence of the conditions of observation (the limits of human perception, the fleeting presence of the viewer in the gallery) than an internal, material property of the biotope. Viewers looking at the biotope another day will see a different "image." Given the cyclical nature of this "image," each "image" seen at a given time is but a moment in the evolution of the work,

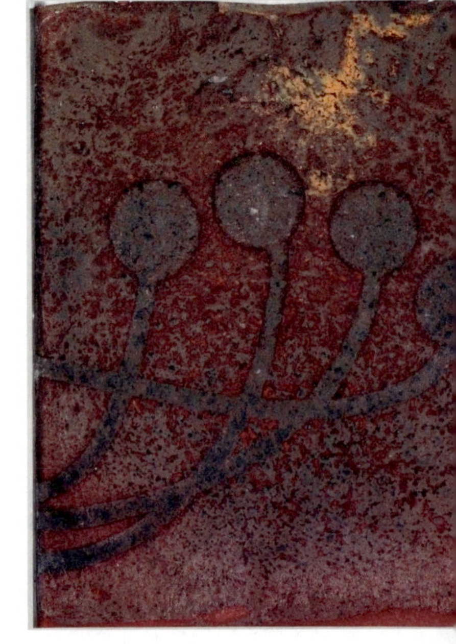

Fig. 35 Eduardo Kac, *Clairvoyance*, biotope, 2006. Collection Parco

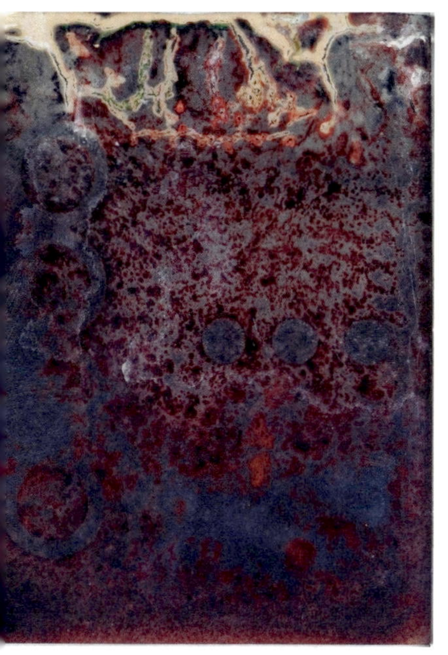

Private collection, Rio de Janeiro. © Eduardo Kac

an ephemeral snapshot of the biotope's metabolic state, a scopic interface for human intimacy.

Each of my biotopes explores what I call "biological time," which is time as it manifests throughout the life cycle of a being itself, *in vivo* (contrary to, say, the frozen time of painting or photography, the montaged time of film or video, or the real time of a telecommunications event). This open process continuously transforms the "image" and may, depending on factors such as lighting conditions and exhibition length, result in its effacement—until the cycle begins again.

The biotope has a cycle that starts when I produce the self-contained body by integrating microorganisms and nutrient-rich media. In the next step, I control the amount of energy the phototrophic microorganisms receive in order to keep some of them active and others in suspended animation. This results in what the viewer may momentarily perceive as a "still image." However, even if the "image" seems "still," the work is constantly evolving and is never physically the same. Only time-lapse video can reveal the transformation that a given biotope undergoes in the course of its slow change and evolution.

To only think of a biotope in terms of microscopic living beings is extremely limiting. While it is also possible to describe a human being in terms of cells, a person is much more than an agglomerate of cells. A person is a whole, not the sum of its parts. We should not confuse our ability to describe a living entity in a given manner (e.g., as an object composed of discrete parts) with the phenomenological consideration of what it is like to be that entity, for that entity. The biotope is a whole. Its presence and overall behavior are that of an artwork that is literally alive.

It is as a whole that the biotope behaves and seeks to satisfy its needs. The biotope asks for light and, occasionally, water. In this sense, it is an artwork that requests the participation of the viewer in the form of personal care. Like a pet, it will provide company and produce more colors in response to the care it receives. Like a plant, it will respond to light. Like a machine, it is programmed to function according to a specific feedback principle (e.g., expose it to the ideal temperature, and it will grow more, but extreme cold or heat will discourage activity). Like an object, it can be boxed and transported. Like an animal with an exoskeleton, it is multicellular, has a fixed bodily structure and is singular. What is a biotope? It is its plural ontological condition that makes it unique. ∞

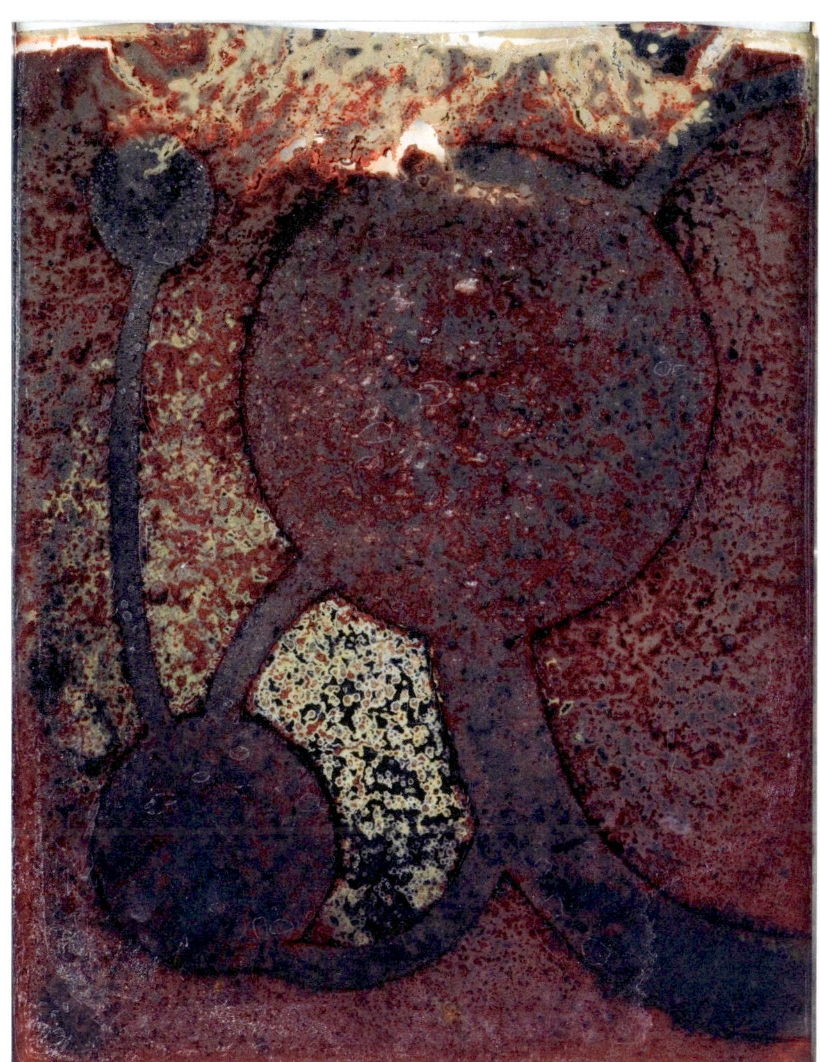

Fig. 37 Eduardo Kac, *Apsides*, biotope, 2006. Private collection, Rio de Janeiro. © Eduardo Kac

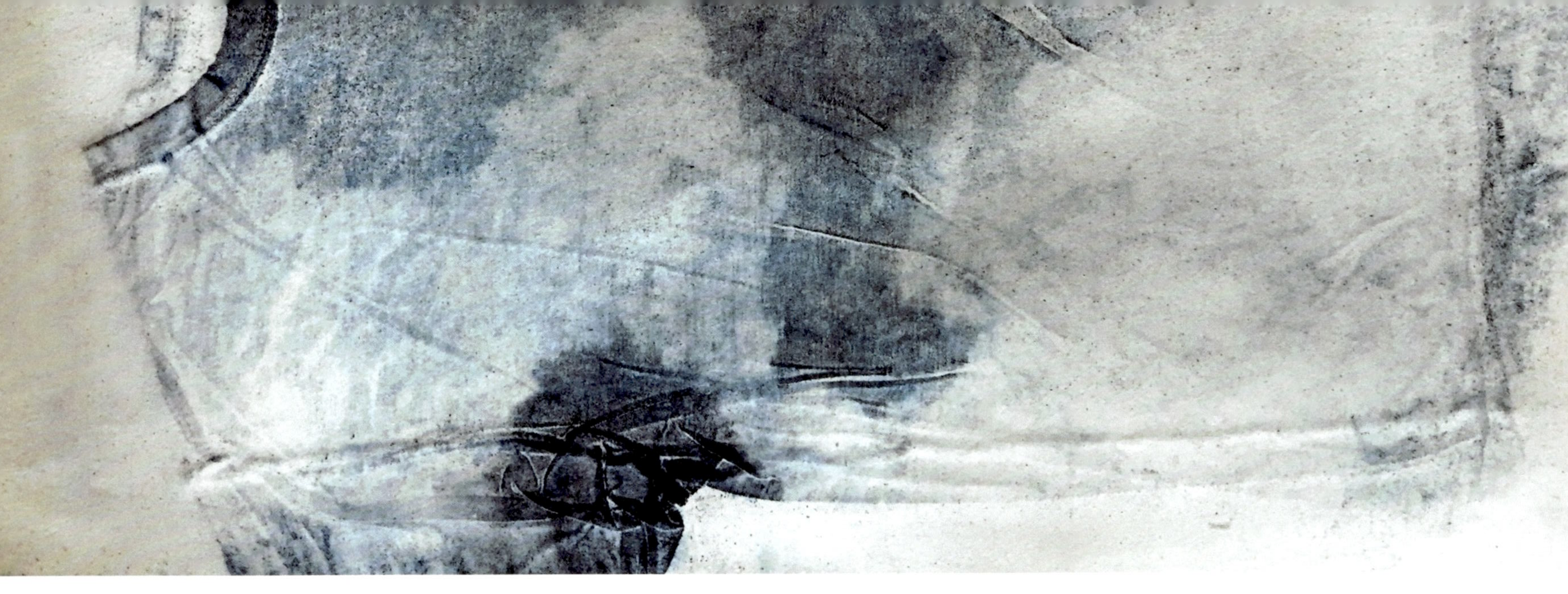

Labor: Work, Difference, and the Post-Anthropocentric Body[1]

Paul Vanouse

Fig. 38 Paul Vanouse, *Self-Blue*, 2017, sweat-stain transfer print. © Paul Vanouse

Introduction

My bio-media artwork of the past twenty-five years has been concerned with how to measure and analyze differences in bodies using biotechnological methods—from anthropometry and physiognomy to contemporary genetics and microbiome research—and with critiquing these supposedly objective techniques and highlighting the biases leading to sociocultural constructions of otherness, such as "race."[2] Organisms like fungi and bacteria may serve as a counterpoint to simplistic distinctions between concepts of same and different, singular and multiple, surface and structure, visible and hidden, and representation and reality.

My use and foregrounding of microbial agents in more recent projects, beginning with *Labor* (2019),[3] has furthered my critique of difference-making from three angles by demonstrating: 1. that the construction of human difference (categorizing and racializing) incorporates nonhuman others to create visceral particularities, which thus should not be attributed to a singular essential human subject type; 2. that these microbes, in that they co-comprise our bodies and our "selves," further complicate any simple self/other dichotomies (individualizing); and, finally, 3. that agency—the question of who is working for whom, in our relationship with microbes (hierarchization)—is a Möbius strip. The gnarly problem of which is on top and which on the bottom has no resolution.

It is in this sense that I describe my approach in *Labor* as *post-anthropocentric*. It is not necessarily post-humanist or anti-humanist, but it is anti-reductive and anti-essentialist. It is a project about point of view, postulating an approach to human-ness, a method of investigation, that foregrounds microbial processes. Here, the post-anthropocentric body serves as a host to

1. This chapter is an abridged and updated version of a text previously published as "Labor: The Post-Anthopocentric Body at Work," in "On Microperformativity," ed. Jens Hauser and Lucie Strecker, *Performance Research* 25, no. 3 (2020): 32–37.

2. Paul Vanouse, *Difference, Sameness and DNA Investigations in Critical Art and Science* (Cham: Palgrave Macmillan, 2024).

3. *Labor* received a Golden Nica in the category Artificial Intelligence & Life Art at the Prix Ars Electronica and was showcased at the Ars Electronica Festival in 2019.

a myriad of flows, forces, metabolisms, behaviors, and perhaps performances. "Our" microbiota, the microbes that liƴe upon or within "our" human tissues and fluids, orchestrate these activities. The work of such nonhuman agents living as "micro performers" on the human epidermis is the focus.

Labor is a dynamic, multi-sensory art installation that endeavors to recreate the scent of human exertion. There are, however, no people involved in making the smell—it is created by bacteria propagating within the artwork's three bioreactors. Each bioreactor incubates a species of human skin bacteria responsible for the primary scent of sweating bodies: *Staphylococcus epidermidis, Corynebacterium xerosis*, or *Propionibacterium avidum*. Human sweat itself is odorless: it is these bacteria feeding upon the components of "our" sweat that create volatile, odiferous chemical compounds that "we" associate with sweat and physical effort. Their scents intermingle in the central glass bell jar, in which a sweatshop icon, a wearer-less white T-shirt, is infused while the scents disseminate out, intensifying throughout the exhibition.[4] The T-shirt, a metonym for the human, is empty. After all, the glass bell jar housing it has historically been a device for extinguishing liƴe. Conversely, the microbes reside, decentered, in inverted bell jars teaming with liƴe and liƴing processes—perhaps even in excess as they generate familiar odiferous compounds. It is the microbes, then, which produce humanness as an effect of their own existence.

Labor reflects upon "our" changing understanding of what "we" are. Microbes in and on the human body vastly outnumber human cells and help regulate many of "our" bodily processes, from digestion and immunity to emotional and physiological responses like sweating. "Our" microbiota are integral to who and what "we" are and complicate any simplistic sense of (an indivisible) self. Likewise, the smell of the perspiring body is not just a human scent, unless "we" are willing to redefine what "we" mean by *human*. In *Labor*, the microorganisms ironically produce the scent of sweat, not as a vulgar *byproduct* of production, like in the

4. Paul Vanouse, "Labor," accessed December 20, 2019, www.paulvanouse.com/labor.html.

factories of the nineteenth and twentieth centuries, but, purposefully, as an evocative *end product*.[5]

Whereas a traditional, anthropocentric worldview considers all activities of, on, and within the human body as unified *human activities*, a contemporary, post-anthropocentric perspective suggests that humans are not only hosts to other organisms, but that these are collaborative, symbiotic agents of "our" human identity.[6] Nearly one hundred years ago, Sigmund Freud dismissed the humanist notion that "we" are rational creatures by showing that "we" are slaves to primal, unconscious drives and desires, thus dethroning mankind from its lofty perch of self-knowledge.[7] More recently, contemporary scholars have challenged "our" ontological assumptions that designate the self as a singular whole, suggesting that humans are merely sentient tubes containing a multitude of intimate microorganisms and metabolic processes.[8] The artistic research that I have carried out in biological laboratories and philosophical libraries, and the resulting artwork, *Labor*, chart how coordinated microbial performances across the human epidermis undermine any simplistic sense of a unified, rational, human *self*.

Our bodies sweat. "Our" microbes stink.

Humans are a rare type of mammal in that they regulate their body temperature through perspiration. Only primates—like monkeys and apes—and horses have a sufficient quantity of sweat glands to allow them to sweat as "we" do. It is the evaporation of this moisture that draws heat from the skin to cool the body. Most others reduce heat through their mouths by panting, or, in the case of the antelope, through a specialized network of tiny heat-exchanging

5. Ibid.

6. Scott F. Gilbert, "Symbiosis as the Way of Eukaryotic Life: The Dependent Co-Origination of the Body," *Journal of Biosciences* 39, no. 2 (2014): 202.

7. Friedel Weinert, *Copernicus, Darwin, and Freud: Revolutions in the History and Philosophy of Science* (Hoboken: John Wiley & Sons, 2008).

8. Lisa Heldke, "It's Chomping All the Way Down: Toward an Ontology of the Human Individual," *The Monist* 101 (2018): 247–60.

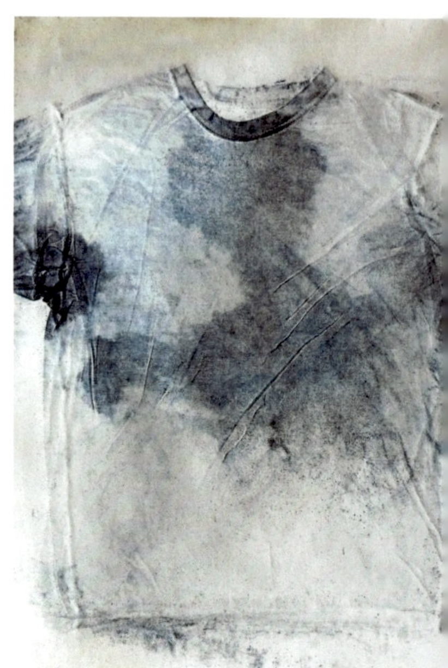

Fig. 39 Paul Vanouse, *Self-Blue*, 2017, sweat-stain transfer print. © Paul Vanouse

veins in the nostrils.[9] Humans possess two main types of sweat glands, eccrine glands and apocrine glands. Eccrine glands cover much of "our" bodies and excrete a watery fluid to regulate body temperature. Apocrine glands are concentrated in "our" warmer, moister regions, like underarms and the groin, and excrete a denser, milky fluid when "we" feel stressed.[10] Apocrine gland quantities vary between individuals and typically begin to develop during adolescence.[11]

When Marx writes of "how man came to be condemned to eat his bread in the sweat of his brow," he alludes to the eccrine sweat of toil and physical exhaustion,[12] whereas when satirist David Sedaris writes, "My conscience is cross-wired with my sweat glands, but there's a short in the system and I break out over things I didn't do … ," he is speaking of the anxious sweat of the apocrine.[13] Zora Neale Hurston's orders to "Sweat, sweat, sweat! Work and sweat, cry and sweat, pray and sweat!" seem to span both bodily secretions, and a multitude of human exploitations as well.[14] In a more recent

9. Nina Jablonski, "The Naked Truth," *Scientific American Special Editions* 22, no. 1 (2012), http://doi.org/10.1038/scientificamericanhuman1112-22.

10. Eugenie Fredrich, Helena Barzantny, Iris Brune, and Andreas Tauch, "Daily Battle Against Body Odor: Towards the Activity of the Axillary Microbiota," *Trends in Microbiology* 21, no. 6 (2013).

11. Katrin Wilke, Annette Martin, Lara Terstegen, and Stefan Biel, "A Short History of Sweat Gland Biology," *International Journal of Cosmetic Science* 29 (2007): 169–79.

12. Karl Marx, *Capital: A Critique of Political Economy—The Process of Capitalist Production, Vol. III—Part II, The Process of Capitalist Production as a Whole* (New York: Cosimo Books, [1867] 2007), 784.

13. David Sedaris, *Dress Your Family in Corduroy and Denim* (New York: Little, Brown and Co., 2014).

14. Zora Neal Hurston and Cheryl Wall, *Sweat* (New Brunswick, New Jersey: Rutgers University Press, [1926] 1997).

example, Marlon James's character, Tracker, discerns between the twin scents thus: "Kava is like most men; he carries two smells. One when the scent runs down back and dries, the sweat of hard work. And one that hides under the arms, between the legs, between the buttocks, what you smell when close enough to touch with lips."[15]

While eccrine sweat is epitomized by the athlete and the manual laborer, apocrine sweat is often *white-collar*—the physical manifestation of emotional stress. Apocrine sweat is particularly unsettling to "our" sense of rational agency as it is beyond "our" control. It is expressed when "our" body feels stressed, fearful, nervous, or anxious, but also when "we" are sexually aroused. It is the highly emotional sweat of psychosomatic reaction, which increases as "we" become conscious of it. It is the sweat that manifests in the office in meetings with one's supervisor, upon cross-examination, when performing onstage, or when making love. It is the bacterial breakdown of this apocrine sweat that produces the most pungent body odors.[16]

Three bacterial genera have been identified as key actors in the production of "our" distinctly human scent, as well as "our" individual scent profiles—*Staphylococcus, Corynebacterium*, and *Propionibacterium*.[17] It is these organisms, "our" microbial symbionts, that liy̸e both upon and within "our" skin, that metabolize "our" varied scentless bodily secretions into "our" particular body odors. *Staphylococcus* is one of the most common and widespread

15. Marlon James, *Black Leopard Red Wolf* (New York, Riverside Press, 2019), 41.

16. Apocrine sweat was also the focus of the work *The Smell of Fear* (2006) by artist Sissel Tolaas, in which she re-created, via chemical synthesis, the scent of men suffering from severe phobias and unable to bear close proximity with other bodies. In this project, Tolaas not only sampled the "smell of fear" and created a powerful olfactory work but also removed unwanted visual baggage by painting the colorless fluid onto a blank gallery wall (I discovered the work exhibited at *sk-interfaces*, curated by Jens Hauser at Casino Forum d'Art Contemporain, Luxembourg, in 2009). While I deeply admire Tolaas's rigorous and ground-breaking work in the area of "olfactory art," throughout the more than twenty years of my own practice, my focus has been on liy̸eness, emergence, and performativity. It is of crucial importance to me that the scents produced in *Labor* are *not* preprepared or artificial, nor mechanically dispensed. It is important to me that this labor is performed liy̸e, it is a living process, and its product is produced far in excess of that which could be created by human toil. *Labor* should make us ponder the very verb "manufacture," as the human here does not *facture*, but is *factured*.

17. Fredrich et al., "Daily Battle against Body Odor."

symbionts of the human epidermis. It transforms leucine and other branch-chain amino acids secreted in sweat into various volatile fatty acids, like isovaleric acid.[18] I describe the resulting smell as "high-pitched" and associate with it with the *permastench* of locker rooms. Others may have different associations, as the human ability to distinguish this scent has been found to vary by over one hundredfold between individuals.[19] *Corynebacterium*, also widespread across the epidermis, alters scentless human steroids and other molecules excreted in sweat to produce malodorous steroids and sulfur compounds. It also incompletely metabolizes long-chain fatty acids, like isosteric acid, into short-chain volatile molecules.[20] It creates odors associated with fear, stress, and anxiety, rather than those resulting from physical activity. Likewise, *Propionibacterium* metabolizes amino acids to produce acetic and propionic acid. These compounds result in what I describe as funky, vinegary, acrid odors. *Propionibacterium* inhabits the moist body regions, both upon and within the human epidermis.

However, all these microbes (and many others) act in complex, often overlapping, communalistic ways on the human body and utilize varied metabolic pathways to produce an aromatic bouquet of a complexity far greater than the sum of their parts. While numerous studies describe the roles that these three bacterial genera play in generating "human" aroma, there is little consensus in defining exactly what each one does, the substrates they act upon, or which of their enzymes are responsible. There appears to be a diversity of capabilities within each genus. For instance, some *in-vitro* studies have found that many *Staphylococcus* species are unable to break down particular carbon-sulfur bonds in precursor molecules, while others have found this ability present

18. Alexander James, John Casey, Della Hyliands, and G. Mycock, "Fatty Acid Metabolism by Cutaneous Bacteria and its Role in Axillary Malodor," *World Journal of Microbiology & Biotechnology* 20 (2004): 787–93; Fredrich et al., "Daily Battle against Body Odor."

19. Idan Menashe, Tatjana Abaffy, Yehudit Hasin, Sivan Goshen, Vered Yahalom, Charles W. Luetje, and Doron Lancet, "Genetic Elucidation of Human Hyperosmia to Isovaleric Acid," *PLOS Biology* 5, no. 11 (2007), https://doi.org/10.1371/journal.pbio.0050284.

20. Fredrich et al., "Daily Battle against Body Odor."

in *Staphylococcus haemolyticus*. Conversely, no singular species of *Corynebacterium* has been found to completely break down human steroidal excretions into the odiferous steroids androstenol and androstenone; rather, the breakdown seems to require a variety of different *Corynebacterium* species acting in concert.[21]

Several scholars have remarked on the profound ontological implications of "our" intimate microbes. Biologist Scott F. Gilbert concludes that "Animals ... cannot be regarded as individuals by anatomical criteria, but rather as holobionts, integrated organisms composed of both host cells and persistent populations of symbionts."[22] Others have redefined biological terms to accommodate the "organism [as] a unit in which all the subunits have evolved to be highly cooperative, with very little conflict."[23] Philosopher Lisa Heldke reflects that, "at the (literal) bodily centre of us, we find not some solid, essential core, but the rest of the world. We are literally tubes full of other organisms. The individual is always already in relation to organisms on the 'outside,' which is always already 'inside.'"[24] "Our" symbiotic relationship with microorganisms confounds "our" ontology of self and other, inside and outside, human and nonhuman.

When "we" scrutinize microbial life and labor across the varying environments of the human epidermis, "we" discover metabolic networks of a complexity that implies a distributed symbiotic metabolism spanning multiple bacterial genera. While there is much to be learned about the majority of bacterial species, studies of common strains provide evidence of multiple metabolic modes that the organisms can utilize depending upon the nutrients and other resources present in their environment.[25] It is likely that "our" varied sweat-eating microbes have overlapping and

21. Ibid.

22. Gilbert, "Symbiosis as the Way of Eukaryotic Life," 202.

23. David C. Queller and Joan E. Strassmann, "Problems of Multi-Species Organisms: Endosymbionts to Holobionts," *Biology and Philosophy* 31 (2016): 855–73.

24. Heldke, "It's Chomping All the Way Down," 248.

25. Kenneth Todar, *Todar's Online Encyclopedia of Bacteriology*, accessed December 20, 2019, www.textbookofbacteriology.net/kt_toc.html.

intersecting metabolisms, and produce differing metabolic enzymes according to the specific nutrients available and the amount of energy required to process them. Thus, like "our" symbiotic selves, the *other* is also impossible to reduce to any stable or simplistic biochemical pathways, or inputs and outputs, but rather contains its own multiplicity. "(M)ultiplicity must not designate a combination of the many and the one," writes Gilles Deleuze, "but rather an organization belonging to the many as such, which has no need whatsoever of unity in order to form a system."[26] "Our" metabolism is largely the result of "our" microbiota, and likewise, their metabolism also contains communalistic relations, of varying forms and terms, with other bacteria, whose metabolisms are partially defined by what "we" feed them.

The Effect of "Our" Scents on Human Behavior and Social Relations

In his introduction to *The Theater of Plautus*, scholar Timothy Moore describes a communication pathway or relay in classical theater: "all communication between playwright and audience is accomplished through the actors."[27] While there is a great deal of nuance to be found in this concise explanation, the role of the actor is essentially that of an intermediary between authorial utterance or intent, and audience reaction or effect. In contemporary science and technology studies, actor-network theory (ANT) complicates the sender-receiver communication paradigm implied in classical theatre, which has continued through modern communication theory. ANT defines the actors involved in creating meaning as both human and nonhuman, material and semiotic, and explains how networks of actors come together to act as a whole.[28]

26. Gilles Deleuze, *Difference and Repetition*, trans. Paul Patton (New York: Columbia University Press, [1968] 1994), 182.

27. Timothy Moore, *The Theater of Plautus: Playing to the Audience* (Austin: University of Texas Press, 1998), 2.

28. Bruno Latour, *Reassembling the Social: An Introduction to Actor-Network-Theory* (Oxford: Oxford University Press, 2005).

The twin significations of the term *actor* here are both in play in the context of scent-creating microbes, which fulfil a highly cultural role in non-verbal cueing and signification. Whereas much of the microbiome/holobiont discussion can be described in ANT terms as a myriad of actors in a complex network *within* a singular human body, microbial odor production serves as a chemical interconnection *between* individual human actors as well (even though this may not be the bacteria's purpose in producing the scent). The phrase "our scents betray us" corroborates the complexity of "our" microbial actors, which signal to an audience, but do not necessarily signify "our" intentions. It is "our" microbiota rather than "our" sweat itself that can signify "our" occupation, "our" comfort level, "our" social status, and the depths of "our" feelings. It may not be an exaggeration to say that "our" microbiota bear primary responsibility for "our" olfactory sensations and experiences of other humans. There are, however, alternative perspectives countering this strong post-anthropocentric claim. Probably the most direct counterargument is the finding of variable proportions of apocrine glands across the human species, and between men and women, pre- and post- adolescent subjects, which could impact the intensity of "our" apocrine odor profile.[29] Arguments like these highlight the human factors (literally gland proportions on "our" bodies) that affect "our" odor.

A more recent counterargument involves the possible genetic basis of human scent profiles and scent perception. In the wake of the human genome project, there has been a strong technological imperative to search for the genetic basis for most human traits.[30] Researchers have found the family of highly polymorphic genes comprising the major histocompatibility complex (MHC) code for the production of non-volatile precursor molecules and have

29. Yoo Hyun Bang et al., "Histopathology of Apocrine Bromhidrosis," *Plastic and Reconstructive Surgery* 98 (1996): 288–92; Wilke et al., "A Short History of Sweat Gland Biology."

30. Much of my bio-media artwork has playfully undermined gene-centric explanations of human differences. For example, *Relative Velocity Inscription Device* (2002) was a liÿe experiment where skin-color genes from my Jamaican/American family members were "raced" against one another in a DNA gel to determine our "genetic fitness."

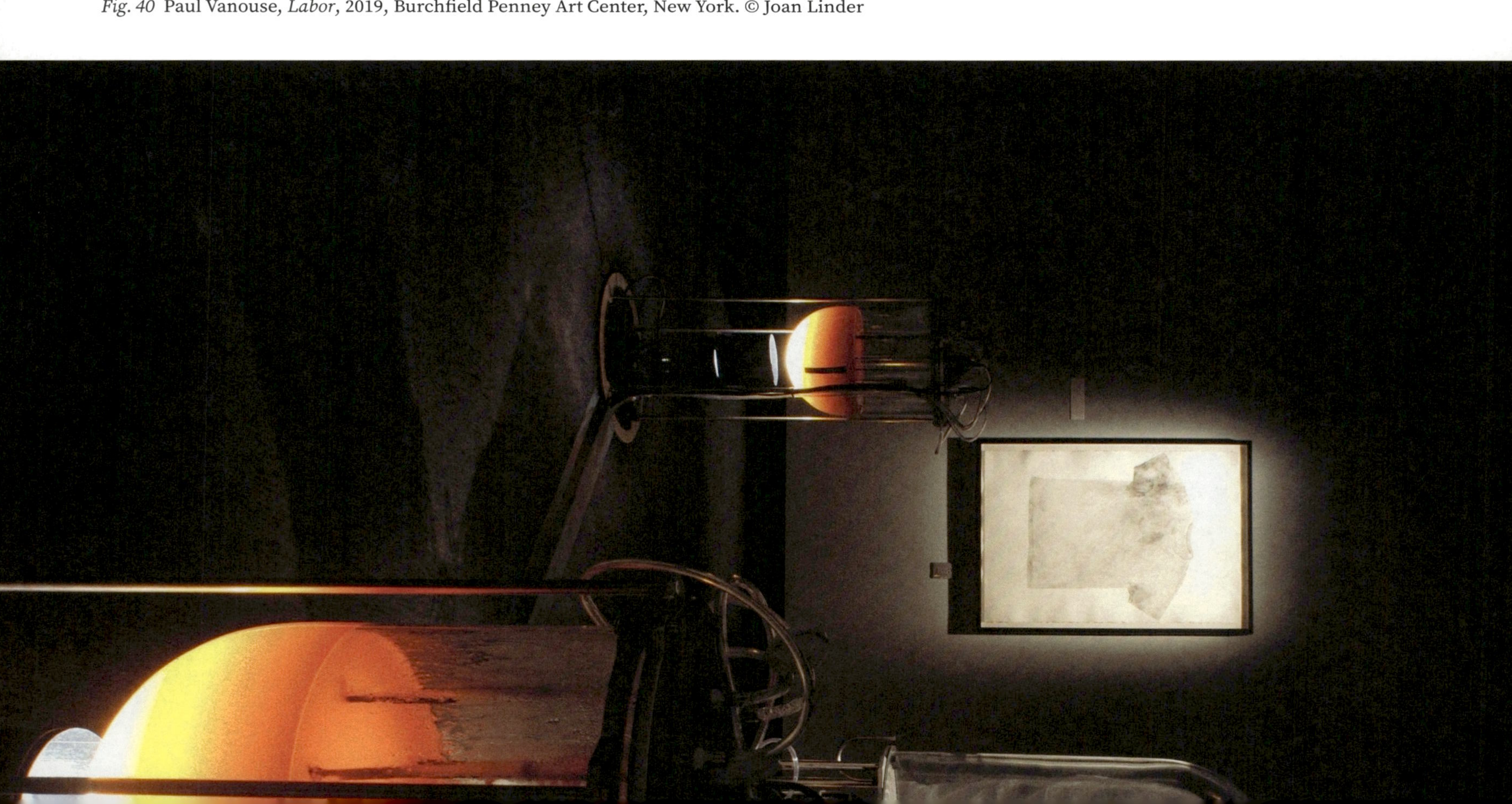

Fig. 40 Paul Vanouse, *Labor*, 2019, Burchfield Penney Art Center, New York. © Joan Linder

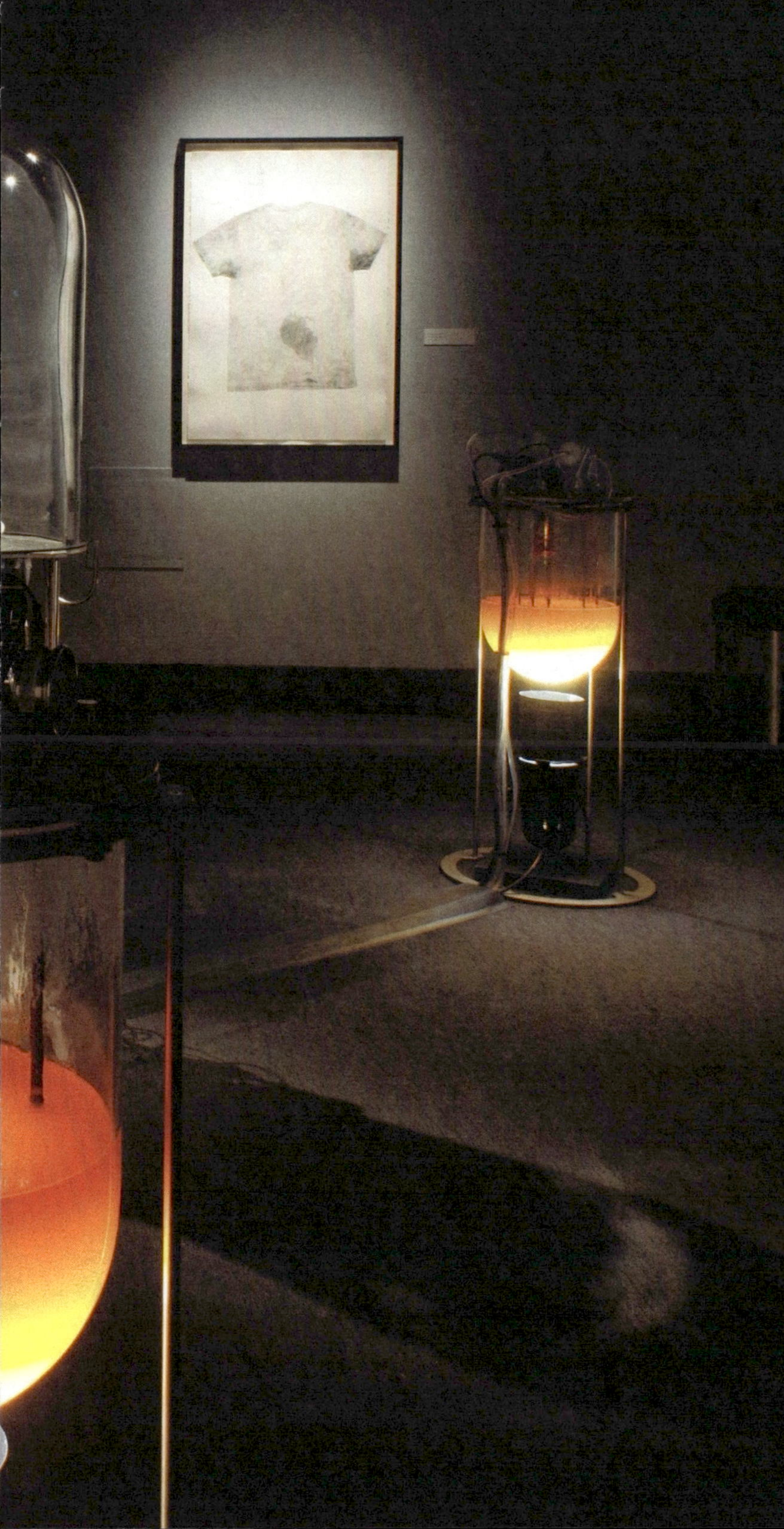

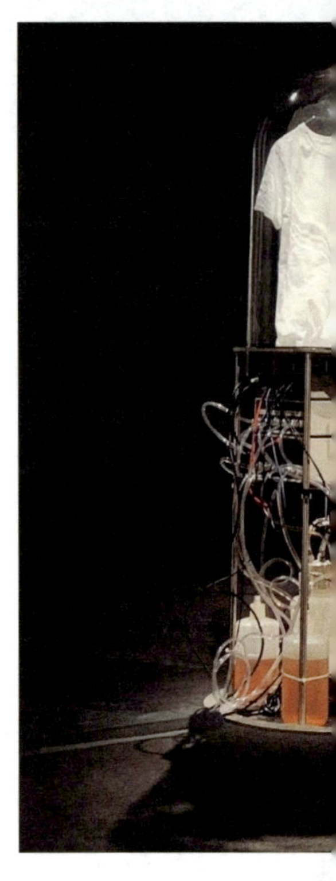

Fig. 41 Paul Vanouse, *Labor*, 2019, Burchfield

shown that these genes vary between individuals.[31] Thus, it could be deduced that, if a person does not possess this particular genetic trait, their body will not produce the proper precursor molecules, and their epidermal microbes will not be able to produce the proper odiferous steroid derivatives. However, there are studies that undermine this genetic explanation by describing how *Corynebacterium* species themselves can mediate the production of these precursor steroids, which play a role in human chemical signaling.[32]

There is a prevalent belief that humans can sense pheromones excreted by one another, which constitute an individual signature as well as a deep, primal key that entices lovers and warns off "our" enemies. While this is a compelling idea, human pheromone receptors, "our" vomeronasal organs, have been proven to be only vestigial, suggesting that what "we" are probably sensing is not pheromones exactly, but complex steroidal derivatives of bacterial manufacture.[33] The pheromone discussion is an appropriate segue to the most profound implications of "our" epidermal microbes. If pheromone receptors are indeed non-functional in humans, the vast majority of the prerational power of pheromones—like the direct connection to "our" reptilian amygdala, the site of "our" most primal urges—can be attributed to "our" bacteria. Bacterial scents might guide "our" deepest, most inexplicable loves and lusts, and "our" most irrational revulsions. What perfumes aspire to do, bacteria have artfully and seamlessly

31. Claus Wedekind, Thomas Seebeck, Florence Bettens, and Alexander Paepke, "The Intensity of Human Body Odors and the MHC: Should We Expect a Link?" *Evolutionary Psychology* 4, no. 1 (2006), https://doi.org/10.1177/147470490600400106.

32. Fredrich et al., "Daily Battle against Body Odor."

33. Didier Trotier, "Vomeronasal Organ and Human Pheromones," *European Annals of Otorhinolaryngology, Head and Neck Diseases* 128, no. 4 (2011): 184–90.

implemented and integrated into each human being. "Our" attractions and "our" disgust, if attributed to bacterial origin, tip the balance of agency from the traditional understanding of "our" relationship as humans with a few microscopic bugs in their bellies to that of emotional puppets that are part of a massive bacterial conspiracy that guides human passions and perhaps even directs human heredity and evolution.

The family tree, the conventional human hereditary pedigree diagram, has, figuratively speaking, always smelled suspicious. It stinks of aristocratic privilege to most of us whose genealogies have been erased through slavery, forced migration, and family separation, cultural erasure and genocide, not to mention poverty and homelessness. And aristocratic hereditary narratives stink of the omissions and polite fictions used to cleanse instances of slave and master co-mingling, class transgression, incest, rape, pregnancy out of wedlock, boy-toys and mistresses, as well as infertility, false pregnancy, and hidden adoption. The family-tree diagram is an ontological diagram of control that rationalizes human genealogy and serves to legitimize privilege. From a post-anthropocentric perspective, the microbial basis of "our" loves and lusts suggests symbiotic bacterial networks are at work in the production of human heredity, of which the family-tree ontology is a cultural byproduct.

Life and Labor as Performance

The *Labor* installation is a post-anthropocentric, scientific theatre in which audiences experience a microbial performance and a simultaneous notable absence. In the middle of the room, and at the absolute center of the installation, connected to a radiating network of tubes, cables, and bioreactors, is an empty, white T-shirt hanging in a glass bell jar. While humans were at the center of "our" worldview, "our" ontology, "our" epistemology, in *Labor* the human has conspicuously vanished. Like Robert Boyle's bell jar, which was central to experiments proving the existence of the vacuum and an icon of the scientific episteme emerging in the seventeenth century,[34] this bell jar rests at the heart of a new, tentative worldview, in which humans have vacated the center.

While viewers search for other visual clues, the scents created by the three bacteria species are intended to create a highly-charged ambivalence among the audience members. These semi-human odors might trigger oppressive or comforting, nostalgic or ominous, amorous or abject emotional responses. *Labor* is apprehended differently than other artworks, as one primary register for perceiving it is olfactory, and the scents produced do not correspond to their expected human subject. As audiences contemplate clues within the installation, they are afforded a window through which to consider many of the issues explored in this paper, such as the concept of human scent created exclusively by nonhumans or—as "we" witness these microbes that live and die to produce and perform humanness—the fatalistic, post-Marxist biopolitical question, "Is life inseparable from labor?"

Labor is not intended to be a new biotechnological simulacrum capable of completely duping audiences. I do not aim, or believe it is possible, to recreate the exact complexity of human aroma using *in-vitro* microbes living in homogeneous liquid cultures. Rather, it is my intent to produce enough aromatic compounds to seem human-like. *Labor* is designed not as a technological end product, but as a tool for post-anthropocentric inquiry. This mode of inquiry also has a particular epistemic aroma, a tincture comprising science experiment, artwork, thought experiment, and philosophical quandary. I have deliberately tried to be backward and oppositional, to reconsider rationally classified relationships: insides and outsides, center and periphery, cause and effect, upside-down and right-side-up. ∞

34. Steven Shapin and Simon Schaffer, *Leviathan and the Air-Pump: Hobbes, Boyle, and the Experimental Life* (Princeton: Princeton University Press, 2011).

Digital Content 05 Paul Vanouse, *Labor*, 2017.

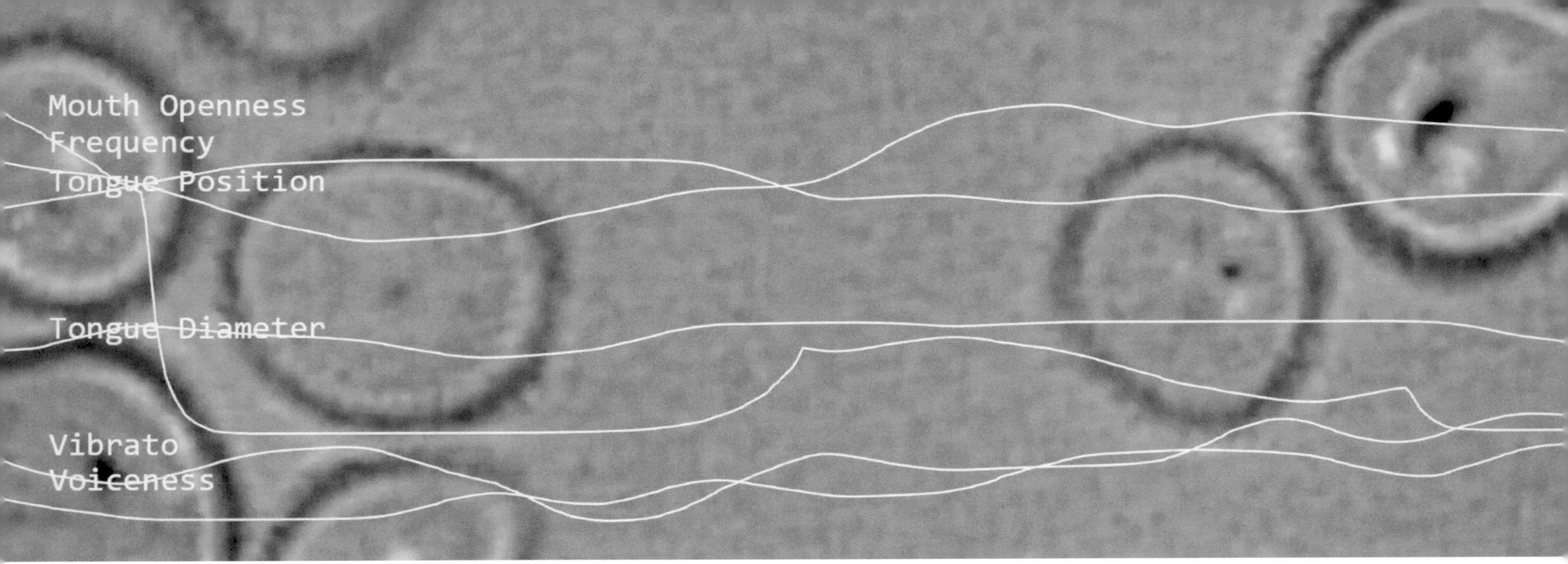

Entangled Voices: How Oral Microbes and Speech Intertwine

Klaus Spiess

Fig. 42 Klaus Spiess, Ulla Rauter, and Emanuel Gollob, *ECOLALIA*, 2020, continuation of *Microbial Keywording*. © Klaus Spiess, Ulla Rauter, and Emanuel Gollob

Escaping linguistic representationalism entirely is undoubtedly a challenge, and there are only few artistic experiments that offer audiences a perspective with which to delve into language as a deeply microperformative, biological process intimately entangled with speech. What insights can be derived from the complex interactions between oral microbes and phonemes? While traditional views like those of John L. Austin and Judith Butler have tended to center on human agency in language, newer frameworks like Karen Barad's concept of "intra-action" are expanding our understanding by highlighting the active involvement of nonhuman elements.[1]

The performance *Microbial Keywording* immersing itself in the tangible materiality of language and speech, promts us to reassess their fundamental nature. Our research focuses on performance practices that confront the entrenched representationalism of language. This idea suggests that linguistic systems often obscure the true material realities they attempt to represent.[2] Representationalism, rooted in a subjective and anthropocentric perspective that can be traced back to Descartes and Kant, interprets the world through linguistic and conceptual frameworks. This approach tends to overlook the dynamic agency inherent in materiality, asserting instead that humans construct the world solely through speech and thought. Recent studies have therefore begun to emphasize the material aspects of language.[3]

Microbial Keywording challenges the notion that humans are the exclusive ontological unit capable of initiating meaningful inquiries into language. We are exploring methods to shift toward a non-anthropocentric ontology that emphasizes nonhuman entities—such as microbes living within us—within a performance context. Referring to Barad's concept of diffraction,[4] our aim is to ele-

1. Karen Barad, *Nature's Queer Performativity* (Durham: Duke University Press, 2011), 125.

2. Karen Barad, *Meeting the Universe Halfway: Quantum Physics and the Entanglement of Matter and Meaning* (Durham: Duke University Press, 2007), 46.

3. See, e.g., Elizabeth de Freitas and Nathalie Sinclair, *Mathematics and the Body: Material Entanglement in the Classroom* (New York: Cambridge University Press, 2014); Jillian R. Cavanaugh and Shakini Shankar, eds., *Language and Materiality* (New York: Cambridge University Press, 2017).

4. Barad, *Meeting the Universe Halfway*, 6.

vate microbes from objects that are merely analyzed and observed to active agencies. These agencies engage with voice, language, and speech, liberating them from human dominance.

Over the past decade, there has been a growing recognition of the profound influence microbes have exerted on human evolution and behavior across various scales.[5] By spotlighting oral microbes as cohosts of human speech, we are pushing back against the traditional notion of language as an exclusively human domain, one that inherently excludes all other species from participation. Our perspective raises intriguing questions for a performative setup: How can we performatively navigate the space between the risk of reducing language to autonomous, detachable phonemes and the joy of expressing the ephemeral qualities of phonemes? To what extent can human speech and arbitrarily motivated words be detached from human origins or principles to incorporate nonhuman agencies? How can we explore the spectrum of vocal and oral modalities to foreground a politics of speech that includes both nonhuman participation and human identity? How can we design a precise microperformative setup that examines the entanglements between nonhumans and spoken language?

In our earlier performances, we had already explored various dimensions of disembodied and embodied speech. In *Spitparty*,[6] we used saliva as a semiotic material, passing it directly from mouth to mouth. *Hare's Blood*[7] featured voicescapes derived from capitalist auction chants,[8] guiding the survival of a living

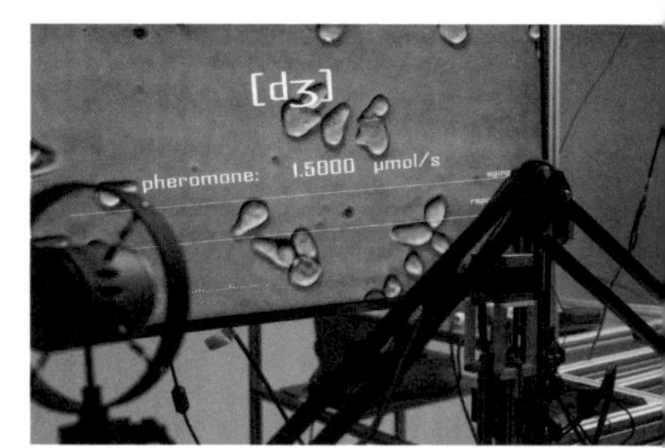

Fig. 43 Klaus Spiess, Ulla Rauter, and Emanuel Gollob, *ECOLALIA*, 2020, continuation

5. Tobias Rees, Thomas Bosch, and Angela E. Douglas, "How the Microbiome Challenges Our Concept of Self," *PLOS Biology* 16, no. 2 (2018): e2005358, https://doi.org/10.1371/journal.pbio.2005358.

6. Klaus Spiess and Lucie Strecker, "Spitparty: DNA Imaging and Aesthetic Mimicry," *Performance Research Journal* 19, no. 4 (2014): 38–44.

7. Klaus Spiess and Lucie Strecker, "In/Valuable Hare's Blood: Performing with Living Relics of Animals," *Performance Research* 22, no. 2 (2017): 115–22.

8. Henry Johnson, "Voicescapes: The (En)Chanting Voice & Its Performance Soundscapes," *Soundscape* 5 (2004): 11.

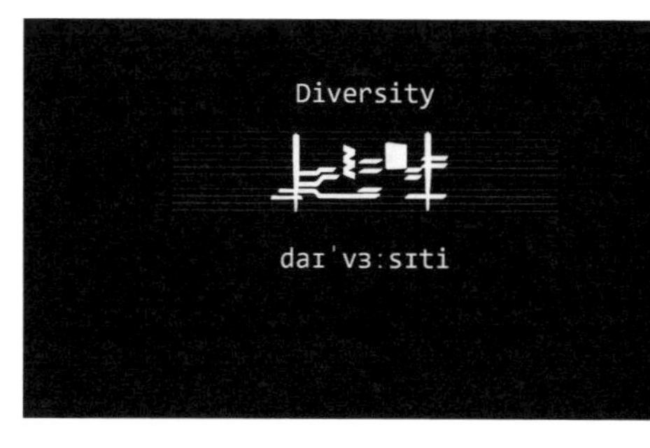

Fig. 44 Klaus Spiess, Ulla Rauter, and Emanuel Gollob, *ECOLALIA*, 2020, continuation of *Microbial Keywording*. © Klaus Spiess, Ulla Rauter, and Emanuel Gollob

microbial artwork. In another recent project, we analyzed the gestures of immunologists as they explained how microbes invade the human body.[9]

Our series *Microbial Keywording* takes these explorations further by focusing on the direct interactions between oral microbes and speech.[10] This series serves as a foundational experiment in understanding how microbes can actively participate in and influence the dynamics of language.

Microbial Coauthorship: Redefining Speech

For our *Microbial Keywording* project, we selected a eukaryotic species inhabiting the human mouth, referred to here as "oral microbes." These microbes are adept at optimizing their environment by anticipating and adapting to changes in chemical and abiotic factors such as oxygen levels, acidity, and temperature.[11] They adjust through processes like cell death, sexual transformation, and reproduction.

The speaker's breath and saliva are essential for the survival of these microbes and the articulation of phonemes. The mouth's specific conditions—oxygen, moisture, and temperature—are closely linked to the rhythmic patterns of salivating and breathing during speech. This unique oral ecosystem forms the foundation for both the microbes' habitat and the vitality of phonemes. The interactions between the speaker's oral environment and microbial activity creates a dynamic setting where microbes actively coauthor speech.

9. Klaus Spiess et al., "Modeling the Immune System with Gestures: A Choreographic View of Embodiment in Science," *Leonardo* 51, no. 5 (2018): 509–16, https://doi.org/10.1162/leon_a_01464.

10. Klaus Spiess and Lucie Strecker, "Microbial Keywording: Gently Whisper to Your Oral Microbes," in *Ars Electronica: Out of the Box*, ed. Hannes Leopoldseder, Christine Schöpf, and Gerfried Stocker (Berlin: Hatje Cantz Verlag, 2019), 80; "Microbial Keywording," *Haus der Kulturen der Welt*, accessed September 4, 2023, https://bit.ly/36ecg8W.

11. Amir Mitchell et al., "Adaptive Prediction of Environmental Changes by Microorganisms," *Nature* 460 (2009): 220–24, https://doi.org/10.1038/nature08112.

To amplify the nuanced variations in saliva flow and acidity,[12] we asked our audiences to repetitively articulate specific phonemes while we collected their saliva. Throughout this process, sensors continuously measured acidity, oxygen levels, and temperature. The microbes' ecological alterations were captured using a dark field microscope and then projected onto a large screen. This set-up created an immersive experience, with the entire performance space resounding with the fluid interplay of voices and microbes.

To enhance the capacity of the audience members' oral microbes to retain the specific phonetic nuances, we employed a meticulously crafted technical setup. This apparatus exposed speakers' microbes to microbial pheromones[13] synchronized with spectrograms of the speakers' phonetic utterances. While the pheromones were eventually taken away again, the microbes retained their highly specific ecological adaptations, affecting rates of reproduction and cell death. This selective influence favored certain input phonemes, thereby reshaping their alphabetical sequence.

On the screen, the data unfolded dynamically, reacting instantly to any changes in the phonetic inputs provided by participants. This interactive mechanism ensured that the visual depiction of microbial activity evolved continuously, reflecting the participants' active involvement with their spoken contributions. In the end, once the microbes had acclimated to the input phonemes, they were reintroduced to some speakers in the form of a probiotic. Each vial was labeled with the corresponding ecologically adaptive phoneme, referred to as a "phonetic mouthwash."

Sensualities of Prelinguistic Articulation

Participants were encouraged to select phonemes from words that held personal significance for them, such as pet names, expressions of emotion, or words introduced to interact with animals

12. James C. Gilchrist and Ernest Furchtgott, "Salivary pH as a Psycho-Physiological Variable," *Psychological Bulletin* 48, no. 3 (1951): 193–210, https://doi.org/10.1037/h0062289.

13. Fabrice Caudron and Yves Barral, "A Super-Assembly of Whi3 Encodes Memory of Deceptive Encounters by Single Cells During Yeast Courtship," *Cell* 155, no. 6 (December 2013), https://doi.org/10.1016/j.cell.2013.10.046.

or children. Through whispered utterances, speakers explored the intimate connection between breath and words described by LaBelle.[14] This rhythmic exploration often led to the nuanced manipulation of articulators and a collective shift from personal to shared vocal expression as interest grew among the audience members.

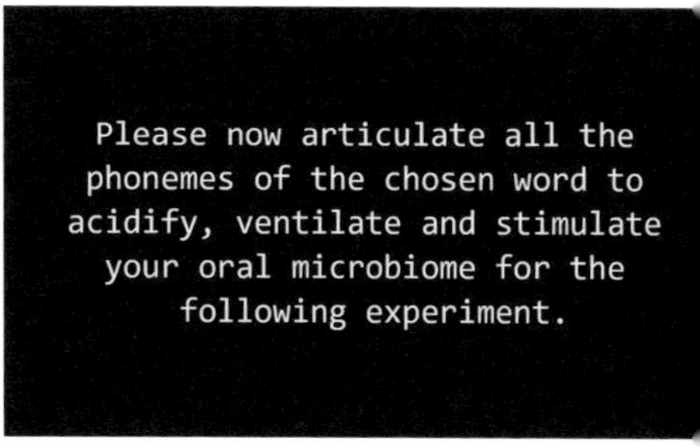

Please now articulate all the phonemes of the chosen word to acidify, ventilate and stimulate your oral microbiome for the following experiment.

Fig. 45 Klaus Spiess, Ulla Rauter, and Emanuel Gollob, *ECOLALIA*, 2020, continuation of *Microbial Keywording*. © Klaus Spiess, Ulla Rauter, and Emanuel Gollob

As participants collectively engaged in babbling, gargling, or sputtering, speech took on a textured, almost uneven quality, where linguistic nuances were swiftly blurred. The evolving rhythms aimed to probe the intricate relationship between phonemes and saliva/air flow, resulting in a form of chanting characterized by specific parameters of pitch, vocal style, and intensity. This chanting often manifested as a vocal soundscape, albeit distinct from traditional singing.

In intimate moments of expression, participants entered what Donald W. Winnicott termed a "transitional phenomenon."[15] Through their salivation, participants appeared to feed on the phonemes they had chosen, akin to an infant "suckling" for comfort. Winnicott viewed this protolanguage as soothing, suggesting that individuals felt they were creating their own phonetic constructs, symbolic of an absent maternal figure or, in our performance, an absent signified.

Michel de Certeau highlights the tension that such utterances place language under, describing them as the pleasures of prelinguistic mouth movements that disrupt the boundaries of meaning and thereby hint at a vision of a post-linguistic future rooted in vocal utopia.[16] As language acquisition progresses, individuals

14. Brandon LaBelle, *Lexicon of the Mouth: Poetics and Politics of Voice and the Oral Imaginary* (New York: Bloomsbury Academic, 2014).

15. Donald Woods Winnicott, *Playing and Reality* (New York: Routledge, 2005), 1–34.

16. Michel de Certeau, "Vocal Utopias: Glossolalias," *Representations* 56 (1996): 33, https://doi.org/10.2307/2928706.

typically lose the ability to distinguish between all the phonemes in human language. However, these utterances served to heighten the audience's awareness of phonemes as representations of the microbes depicted on the screen.

Exploring the Dual Nature of Phonemes

The human voice is a complex interplay of various components: the lungs providing airflow, the vocal folds nestled within the larynx, and the articulators—such as the palate, tongue, cheeks, teeth, lips, and salivary glands—that shape and filter the sound produced by the larynx. To amplify saliva flow and the microbial responses linked to these articulations, we encouraged the audience to physically embody phonemes through gestures. This practice not only heightened participants' awareness of the fact that they were articulating specific phonemes and consonants but also provided them with a tangible connection to the mechanics of speech.

Initially, participants began by employing a repertoire of mouth actions that mirrored certain articulatory, gestural movements of the hands—a phenomenon known as "mouth echo."[17] These gestures were not directly derived from spoken words, nor were they visually motivated; rather, they necessitated specific exhalations or inhalations of breath, often accompanied by changes in mouth configurations during phoneme articulation. The aim was to instill in participants a sense of non-arbitrariness, where hand movements influenced oral actions, rather than the reverse, as commonly observed in gestures accompanying speech.

Our objective was to translate manual actions into oral expressions, transforming ambivalent units of an iconic manual communication system into what appeared to be a largely non-arbitrary, vocal communication system—a transformative experience for participants.

17. Bencie Woll, "Moving from Hand to Mouth: Echo Phonology and the Origins of Language," *Frontiers in Psychology* 5 (2014), https://doi.org/10.3389/fpsyg.2014.00662, 662.

Engaging Microperformatively with Oral Microbes

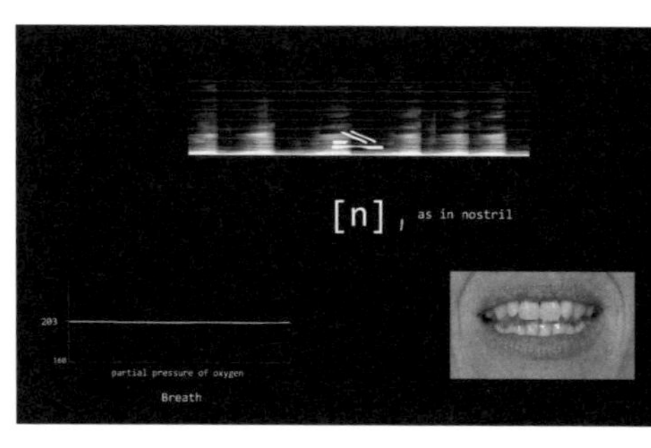

Fig. 46 Klaus Spiess, Ulla Rauter, and Emanuel Gollob, *ECOLALIA*, 2020, continuation of *Microbial Keywording*. © Klaus Spiess, Ulla Rauter, and Emanuel Gollob

The intricate world of oral microbes harbors capacities that go far beyond our conventional understanding. As Nobel laureate Joshua Lederberg has noted, microbes exhibit a level of interactivity surpassing that of humans.[18] Unlike human cultures, they thrive in collectives, reproducing prolifically.[19] Microbes serve as intense intermediaries between humans and their surrounding ecosystems.[20] Their metabolisms and reproductive processes respond sensitively to environmental changes, including those driven by industrial activities, leading to rapid shifts in microbial community dynamics.

When faced with scarcity, microbes adapt by altering their metabolic and sexual behaviors, transitioning from energy-saving asexual reproduction to sexual replication in order to ensure genetic diversity. This sexual system is not merely a biological function—it is deeply ingrained in their existence. Oral microbes optimize their sexuality to maximize their perception of the oral environment. In times of famine, they signal for parts of their community to perish, recycling cellular waste in order to sustain the collective and ensure survival. Microbes have evolved sophisticated means of attracting and transporting species through volatile compounds, making them true cosmopolitans. For microbes, circulation is synonymous with survival—a principle shared with language.

In *Microbial Keywording*, this microbial ecology profoundly influences, and is influenced by, the performance. Oral microbes perceive phonemes as integral to their ecological existence, favoring certain sounds while rejecting others. Substance and form,

18. Joshua Lederberg, “The Microbe’s Contribution to Biology—50 Years After,” *International Microbiology* 9, no. 3 (2006): 155–56.

19. Dominika M. Wloch-Salamon, Roberta M. Fisher, and Brigitte Regenberg, “Division of Labour in the Yeast: Saccharomyces Yerevisiae,” *Yeast* 34, no. 10 (2017): 399–406, https://doi.org/10.1002/yea.3241.

20. Pietro Buzzini, Marc-André Lachance, and Andrey Yurkov, eds., *Yeasts in Natural Ecosystems: Ecology* (Heidelberg: Springer, 2017).

materiality and intelligibility, converge in the world of microbes. Rather than being constrained by fixed metabolic or sexual boundaries, they embody the mutability of matter, engaging in a dynamic relationship with their environment.

Microbial Keywording explores the nuanced pitches and tones that create varied living conditions for these microbes. By relinquishing the exclusive ownership of voice and speech to microbes—nonhuman entities residing and acting within the human body—the human voice and speech transform into essential materialistic processes, constituting a materialist practice.

We refer to this intricate interplay between devices, microbes, saliva chemistry, voices, and words as an "ecomaterial" or "ecolinguistic" realm—a realm where ecological patterns emerge through mutual interaction and interpretation.

Materializing Memory

In the realm of *Microbial Keywording*, we challenge the traditional ontology of language and speech, moving away from the notion of an individual rational "I" toward a materialist discursive practice, where microbes play a pivotal role. Here, speech emerges not solely from the essence of the human speaker but also from the dynamic needs of the nonhuman entities dwelling within the speaker's mouth. Human speech undergoes fragmentation, transforming into an ecological amalgam of nonhuman and human elements—a disarticulated organism comprised of diverse forces, needs, thresholds, and intensities.[21] This voice lacks a singular subject, transcending the humancentric perspective that typically guides ontological inquiry. Instead, it becomes a process of interconnecting different bodies, spaces, times, expressions, and beings, giving rise to a non-essentialist layer of oral ecological craving and desire.

We conceive of voice as a recording surface that captures the entire process/performance, including microbial adaptations, rather than as a singular process aimed at translating subjective

21. Gilles Deleuze and Félix Guattari, *A Thousand Plateaus*, trans. Brian Massumi (London: Continuum, [1980] 1987), 177.

experiences into a unified and individualized voice.[22] In *Microbial Keywording,* microbes transcend the boundaries between organisms even before the act of recording. Rather than archiving the voice as a static object in space and time, these microbial entities record a constantly evolving, dehierarchized, sexually reproductive, and metabolically transformative aspect of voice.

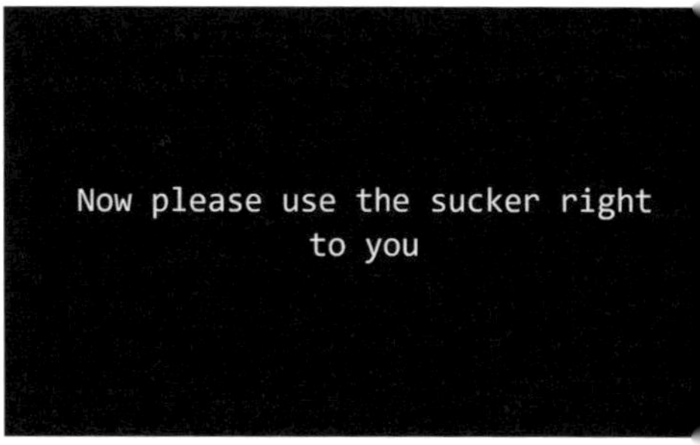

The notion of breaking away from the confines of the humanist subject hinges on the concept of memory. While traditional memory is often viewed as static, limited to the cognitive experiences of the knowing subject, microbial memory in *Microbial Keywording* presents a different paradigm. It constitutes an ecological, materialist memory, reacting directly to its environment within the confines of its cellular membrane. The voice thus embodies a shift toward a different ontology—a microbial ontology—where memory encapsulates not only the individual cognitive past of its human host but also collective environmental history. This memory encompasses li*f*e itself, as Gilles Deleuze suggests, resisting specification or attachment to a singular self and transcending the limited "experience" of individual subjects.[23]

Fig. 47 Klaus Spiess, Ulla Rauter, and Emanuel Gollob, *ECOLALIA*, 2020, continuation of *Microbial Keywording.* © Klaus Spiess, Ulla Rauter, and Emanuel Gollob

Microbes, Voiced

In *Microbial Keywording,* phonetically entangled microbes became part of a conventional biological recording process in the Petri dish. This ability for infinite replication altered traditional notions of vocal authenticity. While the experience of hearing

22. Gilles Deleuze and Félix Guattari, *Anti-Oedipus: Capitalism and Schizophrenia,* trans. Robert Hurley, Mark Seem, and Helen R. Lane (Minneapolis: University of Minnesota Press, [1972] 1983), 11.

23. Gilles Deleuze, *Pure Immanence: Essays on A Life,* trans. A. Boyman (New York: Zone Books, [1995] 2005).

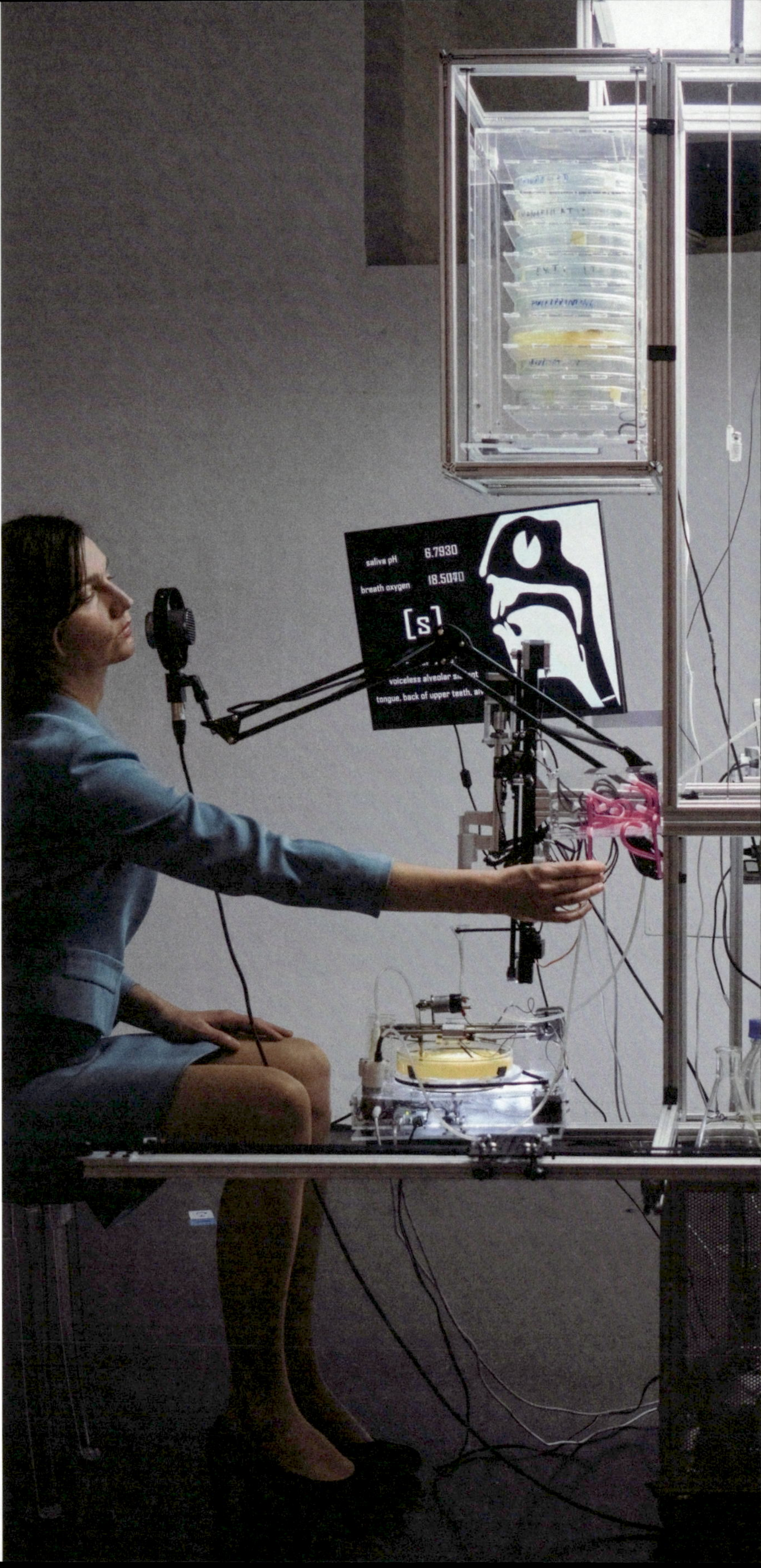

Fig. 48 Klaus Spiess and Lucie Strecker, *Microbial Keywording*, 2020, exhibition view, MuseumsQuartier Vienna, Showroom D. © David Pujadas Bosch

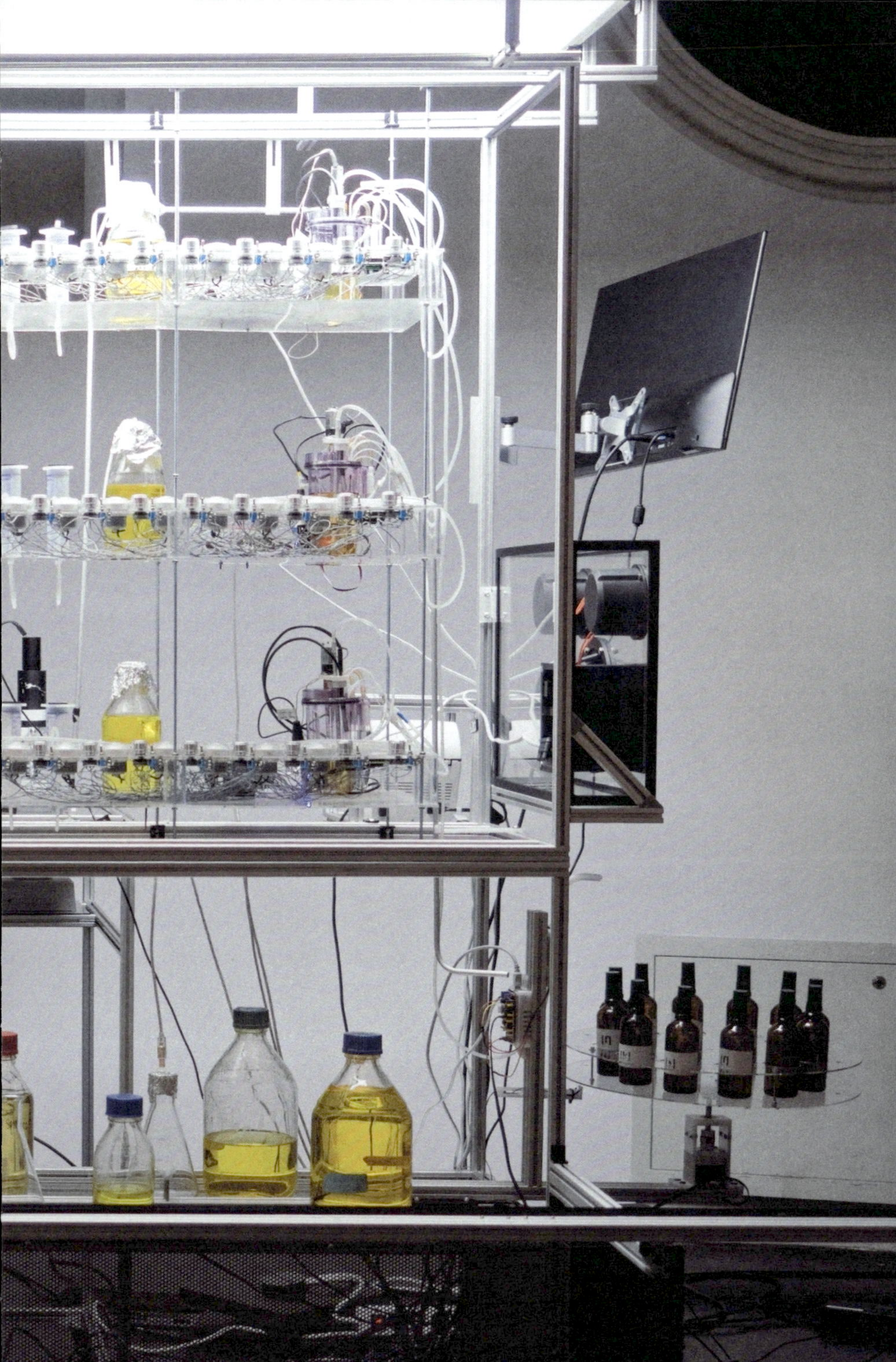

stories "passed on from mouth to mouth"[24] is highly personal, the biological recording in *Microbial Keywording* led to an experience of disembodiment. The recording depended on microbes being transferred while speaking, severing the physical link between the speaker and their unique voice. As biological reproduction and cell death dictated the longevity of the voice, it became reminiscent of a living epitaph.

Microbial Keywording allowed us to reconsider the dynamics and actions that take place in our mouths within their own biological system. What was once seen as an authentic, creating subject is now merely a cohost, part of an intra-action that produces language. Furthermore, reflecting on language involves considering the ownership of that language. Who is speaking? Whose voice does the audience hear? Where does it originate? *Microbial Keywording* revealed that these questions are complex and challenging to answer. It closely resembles a "voice without a subject,"[25] a concept Mazzei (2016) draws from Deleuze and Félix Guattari's "body without organs."[26]

The term "body without organs" (BWO) originated in Antonin Artaud's proposition that removing the organs from the body to eliminate all automatic reactions would ultimately liberate the body.[27] Artaud's vision was that of a theater of gestures, sounds, and screams that would transcend and obliterate language, thereby erasing known signifiers. For Deleuze, Artaud and his Theater of Cruelty exemplified the schizophrenic who, uninterested in social norms or conventional language, delves deeper into the materiality of existence.[28] However, Artaud's concept of performance remained humancentric, seeking catharsis. To truly embrace the BWO

24. Walter Benjamin, "The Storyteller: Reflections on the Work of Nikolai Leskov," in *Illuminations*, ed. Hannah Arendt, trans. Harry Zohn (New York: Schocken, 1968), 84.

25. Lisa A. Mazzei, "Voice Without a Subject," *Cultural Studies ↔ Critical Methodologies* 16, no. 2 (2016): 151–61, https://doi.org/10.1177/1532708616636893.

26. Deleuze and Guattari, *A Thousand Plateaus*, 149–66.

27. Antonin Artaud, "To Have Done with the Judgment of God: An Approximation in English," trans. Guy Wernham, *Northwest Review* 6, no. 4 (1963): 61.

28. Gilles Deleuze, *The Logic of Sense*, trans. Mark Lester (New York: Columbia University Press, 1990).

concept would mean recognizing the body as an assemblage of forces and nonhuman agents such as microbes, which thrive, intra-act, and take control of the body.

What are the forces and agents that constitute voice in *Microbial Keywording*? They are microbes, pheromones, and the flow of saliva. Through this reduction the voice is liberated, aligning with Artaud's vision. It is freed from the confines of the human subject. Artaud described this as "cruel" as it abandons established humanist patterns. Our performance aligns with Artaud's concept of a Theatre of Cruelty. In *Microbial Keywording*, microbes—not humans—are the main characters, replacing the human-centered view of body and voice with dynamic interactions between microbes and matter.

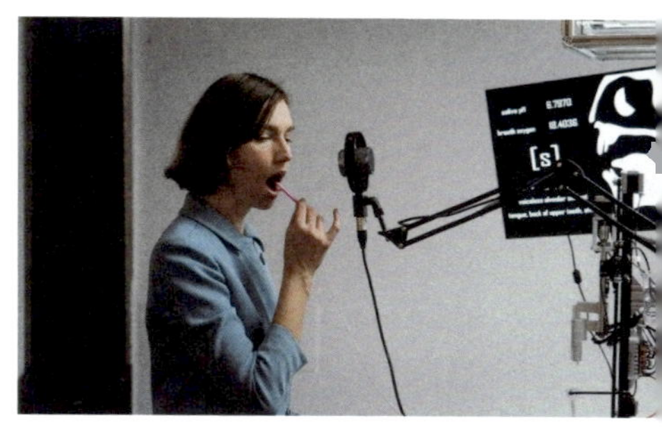

Fig. 49 Klaus Spiess and Lucie Strecker, *Microbial Keywording*, 2020, exhibition view, MuseumsQuartier Vienna, Showroom D. © David Pujadas Bosch

The concept of the BWO transforms into a *body with microbes*—a body teeming with sexually active, transforming, migrating, ever-changing microorganisms. These microbes inhabit all organs and significantly influence their functions. However, the term "body with microbes" still positions the human body as the main actor. In reality, microbes themselves act as a new kind of organ, essential to the human body's functioning and its other organs. Instead of being the primary organ for microbes, the human body serves as a host in which microbes showcase their diverse capabilities. This perspective represents a shift from seeing the body merely without organs or with microbes to viewing it as *microbes with a body*. Similarly, it moves from the idea of a voice without a subject to *microbes with a voice*.

A Diffractive Perspective

In the *Microbial Keywording* performance, microbes intra-act with each other as phonemes, and phonemes intra-act as microbes. Microbes function as a speech apparatus, and phonemes are not just abstract linguistic concepts but living entities. They are constantly active, influencing and changing each other. For instance, microbes can alter their sexuality in response to specific phonetic cues.

The technical equipment used in *Microbial Keywording* serves as a crucial tool for creating diffraction. Phenomena arise from the intra-action between this equipment and the microbes. The equipment is not just used to observe objectively; it is a constitutive part of the performance with the microbes. These phenomena, the smallest units of analysis, are not simply observed by the instruments but actually emerge within the intra-action between the equipment and the microbes.

It is not possible to observe a phenomenon using analytical instruments because the phenomenon does not exist before the experiment; it arises out of it Nothing passively waits to be observed; everything emerges from the intra-actions between the instrument and the object. This includes intra-actions between phoneme improvisation, voice spectrograms, Petri dishes, bioreactors, microbial adaptation, and phonetic probiotics. The diffractive investigative method does not just separate microbes from the human mouth; it combines and separates them simultaneously. This approach highlights the potential intra-actions between microbes in a Petri dish and the phonemes we speak.

First, the microbes are taken from their natural habitat—the human mouth. This separation reveals how interconnected microbes and phonemes are, with the human mouth the area in which this connection takes place. By reading diffractively, we see that the intra-action of speech, microbes, and the technical equipment in *Microbial Keywording* is not just an experimental setup to understand the "true nature" of microbes or speech. Instead, it becomes a performance where new relationships between phonemes and microbes develop and emerge.

As Barad notes, even when the brittlestar loses a limb, it remains connected to its environment.[29] Barad questions whether the separated limb is part of the brittlestar or the environment. Connectivity does not always require physical closeness Similarly,

29. Karen Barad, "Invertebrate Visions: Diffractions of the Brittlestar," in *The Multispecies Salon*, ed. Eben Kirksey (Durham: Duke University Press, 2014), 221–241.

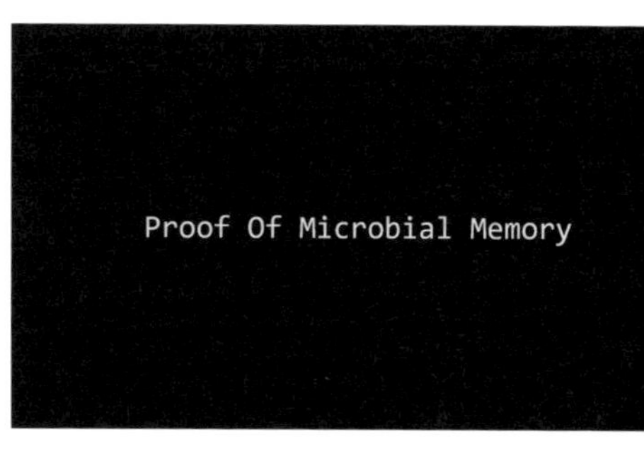

Fig. 50 Klaus Spiess, Ulla Rauter, and Emanuel Gollob, *ECOLALIA*, 2020, continuation of *Microbial Keywording*. © Klaus Spiess, Ulla Rauter, and Emanuel Gollob

microbes are not just part of their surroundings, people, or voices; all these entities are interconnected through their intra-actions. These connections are materially staged and involve discursive practices such as speech. Microbes act as an "externality within," bridging the gap and allowing us to perceive human language in its material form, thereby changing our relationship with it.

Perceiving Language Speak Itself

In today's world, with environmental crises like climate change looming large, Timothy Morton urges us to reconsider our relationship with nature, viewing it as an intimate process intricately woven into our existence rather than as something external.[30] This perspective prompts us to recognize the profound intimacy of microbes in *Microbial Keywording*—where microbial behaviors, such as the way they alter their sexuality within our own mouths, highlight their close connection to us.

While we might not be able to fully escape linguistic representationalism, *Microbial Keywording* allows audiences to reconsider language and speech as processes of intimacy. What can we learn about our speech from the interactions between microbes and phonemes at the heart of our performance? Philosophers like Austin and Butler have traditionally viewed language as performative, but such ideas often focus solely on human subjects. Newer ideas like Barad's concept of intra-action shift the focus to the active role played by objects, rendering research not just a series of measurements but a performance. *Microbial Keywording* goes further by revealing the materiality of language and speech, prompting questions about their ontological status.

Language, speech, and voice transcend mere human constructs—they are deeply intertwined with their material

30. Timothy Morton, "Queer Ecology," *PMLA* 125, no. 2 (2010): 274.

components and the broader world of objects. This perspective allows us to perceive voice as a distinct ontological entity, pulsating with life, evolution, and independence from any single subject. Literature has long hinted at this notion, from the vivid imagery of frozen words in François Rabelais's writings[31] to the enigmatic utterances found in Conrad's *Heart of Darkness*[32], suggesting that language has its own autonomy from human authorship.

Despite these literary insights, conventional depictions often confine language within the human. Our performance, however, breaks free from such constrains, offering participants an immersive exploration of language as something not dictated by humans but orchestrated by microbes—the very building blocks of speech itself. It is akin to witnessing language unfold autonomously, with microbes shaping phonetic patterns and utilizing humans as conduits through which to manifest their preferred auditory environments.[33] ∞

31. François Rabelais, *Gargantua and Pantagruel* (Moscow: Dodo Press, 2008).

32. Joseph Conrad, *Heart of Darkness* (Scotts Valley: CreateSpace Independent Publishing Platform, 2014), 12.

33. Acknowledgements: Funded by the Austrian Science Fund (FWF) AR 687 and V501–G24, the Medical University of Vienna, and the University of Applied Arts Vienna.

Digital Content 06 Klaus Spiess, Ulla Rauter, Emanuel Gollob, and Lucie Strecker, *ECOLAIA*, 2020.

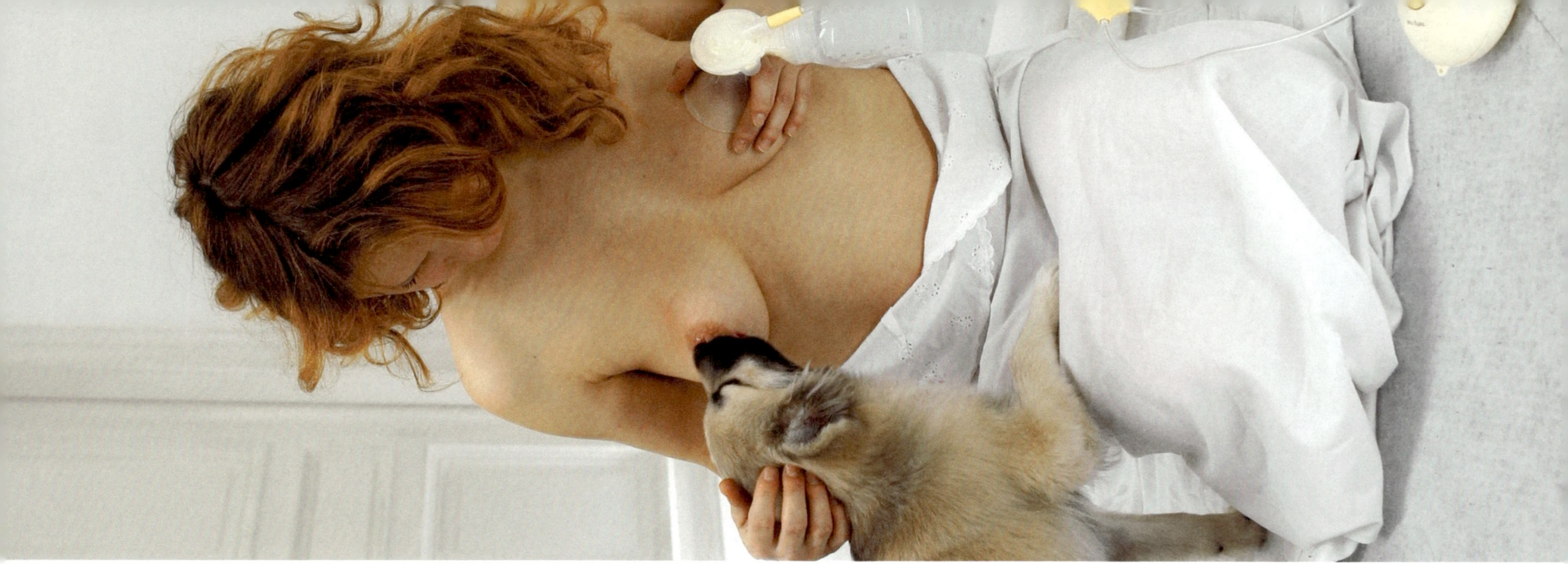

HYBRID FAMILY: An Exchange on the Possibilities of mOtherhood

Jens Hauser and Maja Smrekar

Fig. 51 Maja Smrekar and Manuel Vason, *K-9_topology: HYBRID FAMILY*, 2016, a photo performance produced by Kapelica Gallery, Slovenia, and Freies Museum Berlin, Germany. © Maja Smrekar and Manuel Vason

During a three-month period of seclusion, this durational performance process explored, socially and physiologically, the coevolution and coexistence of humans and dogs as well as possible scenarios of their cross-species hybridization in order to form a hybrid family. In *K-9_topology*, Maja Smrekar's process of *becoming-animal* involved stimulating her pituitary glands by systematically breast-pumping in order to release the hormone prolactin. She also ate a diet rich in galactagogues to promote lactation so that she could breastfeed and raise her newly adopted puppy, and future artistic collaborator, Ada. Overcoming the social and ideological roles of a woman's body and breastfeeding practices, as well as the distinction between private life and political action, the staging and sharing of hybrid family life in an apartment located at Freies Museum Berlin was intended to generate empathy while increasing resistance to the widespread cynicism of the current zeitgeist. Throughout the process, Smrekar engaged in an epistolary exchange with cocurator Jens Hauser between November 2015 and July 2016.

Berlin, November 11, 2015
History of Tears

Dear Jens,

I am excited to start an exchange with you as my dialogue partner throughout this ambitious process, and I anticipate that the following months will be a challenging journey for me. As an artist who observes, digests, and comments on the complex dynamics of our contemporary world, I need to emancipate myself from the primal patriarchal horde, which has entrenched the most horrific neoliberal and capitalistic patterns deep within our societies and cultures, asking myself what sort of "live arts" may indeed be up to meeting the challenges of our damaged planet. How can we maintain empathy for each other in our capitalist ruins, and how can I particularize my own mOtherness in such contexts? Answering these questions means expanding my views beyond anthropocentrism and making my artistic statement on a micro-political level, even a molecular

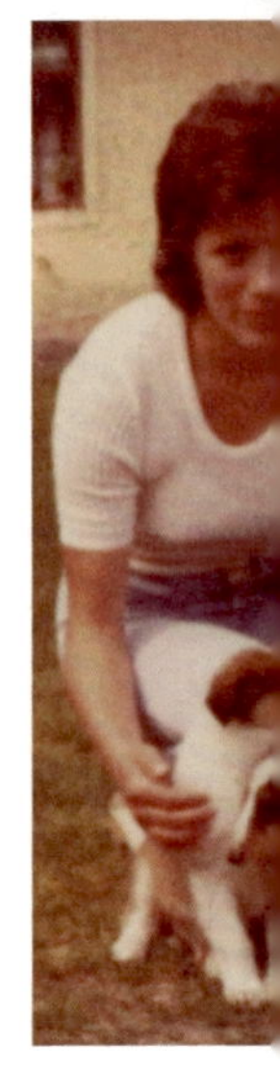

one, adopting a microperformative method to foster empathy for the nonhuman other.

Since the core intention of the whole *K-9_topology* series emerged from the emotional memory of my childhood and youth, through the psychodynamics of my family and, at the same time, my perception of home, I feel it is appropriate to start with my memories of how I felt growing up in my primary family. My parents worked in a leather and fur textile business, located in our house. Therefore, the olfactory memory of leather and fur shaped the essential sensorial memories engraved into my early neural networks on a daily basis. Besides running their business all over former Yugoslavia and strongly delving into the field of cynology, my father was a very keen and active wildlife hunter, a winemaker, and a military reserve officer. My mother was a seamstress and a companion on all of my father's endeavors. Nevertheless, all of this scenery disappeared behind our family's main focus: dogs, the only agents that enabled the three of us to share real, gentle affection with each other.

Around the year 2000, liberal capitalism struck hard into the nascent Slovenian economy. My parents lost their business, house, cars, forests, fields, vineyard due to several mortgages they had taken out in the 1990s while trying to keep the illusion of the easy socialist living of the past artificially aliye. As a consequence, my father committed suicide. All the dogs were either sold, donated, or escaped into the unknown—at least, that was the version the rest of my family told me. I never believed either those stories or those of the other dogs "disappearing." The dogs would usually stay with my family for a couple of years, then I would come home from school one day, and the dog would have gone to an unknown location. The bonds of kinship would be abruptly broken, in addition to the pain and void worsened by the lack of truth, until I mitigated them with new kinship ... with a new dog. Dogs were one of my father's compulsive projects. But since his suicide, I have never again stepped into our house, which was seized by the bank. Our whole property, at the very border of the "Schengen Area," vanished, was erased, as

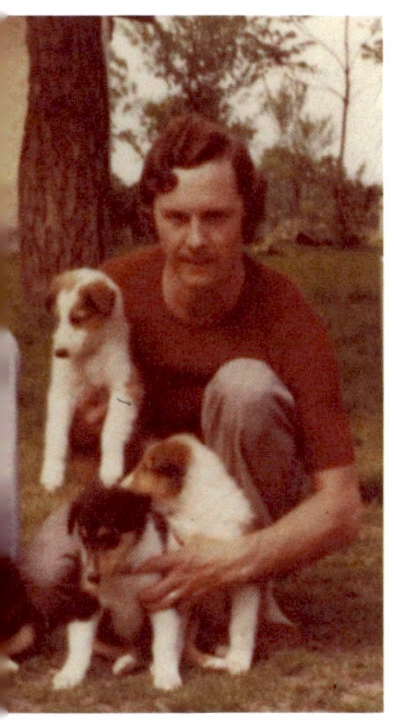

Fig. 52 Una's litter and Smrekar's parents Marija and Branko Smrekar in their backyard, Dobova, Slovenia, 1977. © Unknown

if it had been fenced in with the invisible barbed wire of dark emotions, which will not let me step into my home ever again. Consequently, I started to feel "minor," like an economic migrant searching for their secondary home ever since, feeling deterritorialized, like a kind of a hybrid between a non-native and yet privileged EU citizen at the same time, without fully becoming either. Today, a dear childhood friend of mine sent me a message on Facebook: "Maja, can you imagine? The very place where we spent all our childhood summers by the Sotla river, at the 'green' Croatian-Slovenian border, has been barbed-wired!" But I have been feeling like it has been barbed-wired for fourteen years. I am jealous of all those people in my village who were smart enough and able to keep their homes during the East-West transition in the 1990s and feel ashamed.

Against this background, I feel the need to start a new family. But it needs to *become* a family through deterritorialization! And therefore, it needs to be a HYBRID one. By locating myself within the *becoming-animal* discourse, the whole process of the *K-9_topology* series, especially the *HYBRID FAMILY,* thus constitutes a complex journey for me as an artist who feels the need to observe the complex contemporary dynamics of the external as well as the internal world, and to transpose them into the very molecular depths of myself. For my previous performance *I Hunt Nature, and Culture Hunts Me* (2014),[1] I had somehow already anticipated what now will follow:

> The management of my own animality is still human, but transcendence into a human animal is what I find truly fulfilling. I don't have siblings. I am alone. I do not feel any emotional security. We compose a hybrid family, a subversive cartel. You and I are hunted by those who lack the capacity to shift into the intimacy of companionship beyond the

1. Maja Smrekar (Ewen Chardronnet), "I Hunt Nature and Culture Hunts Me—Rencontres Bandits-Mages 2014," *Vimeo,* 2022, video, 9:14, www.web.archive.org/web/20170815031942/https:/vimeo.com/112481726.

Fig. 56 Smrekar with Doca in the backyard of their family home, Dobova, Slovenia, 1980. © Unknown

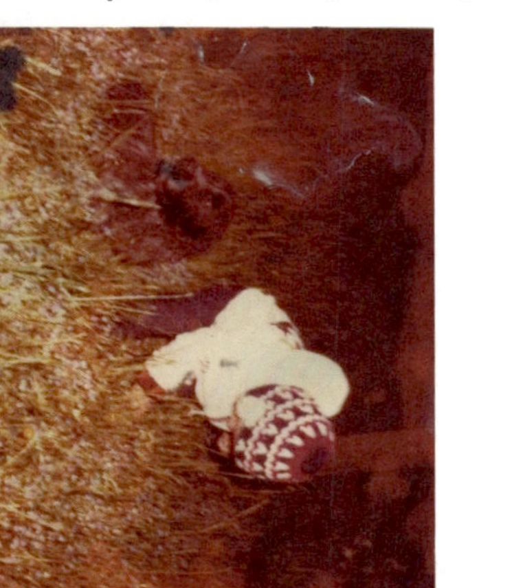

Fig. 53 Smrekar with Doca and her mother Marija in the back-yard of their family home, Dobova, Slovenia, 1980. © Unknown

Fig. 55 Bas and Maja in front of the Smrekar's family home, Dobova, Slovenia, 1984. © Branko Smrekar

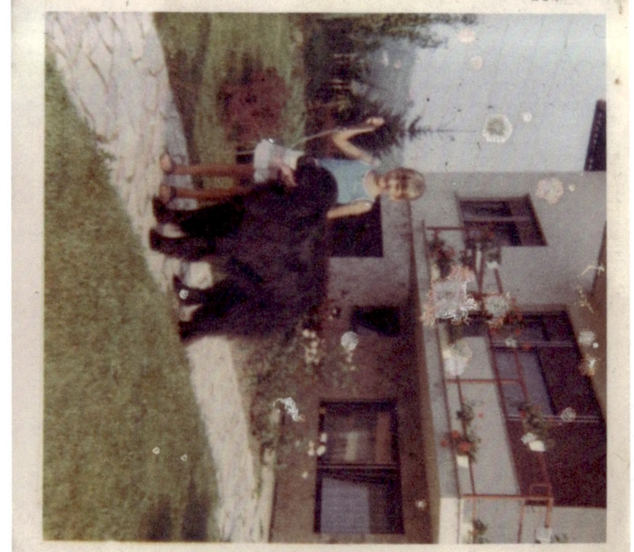

Fig. 54 Nina, Dobova, Slovenia, 1991. © Maja Smrekar

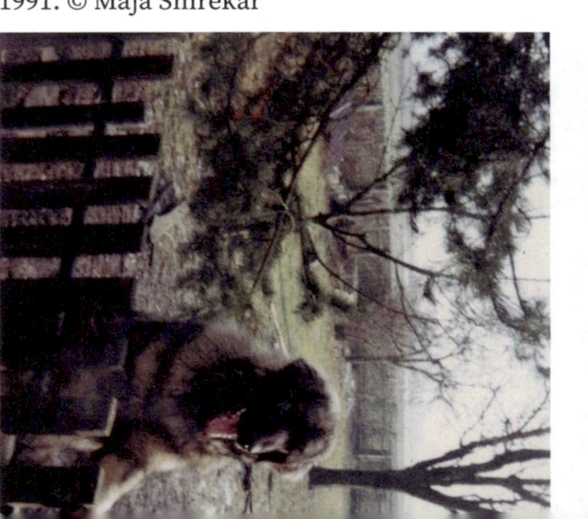

Fig. 57 Aika, Dobova, Slovenia, 2001. © Maja Smrekar

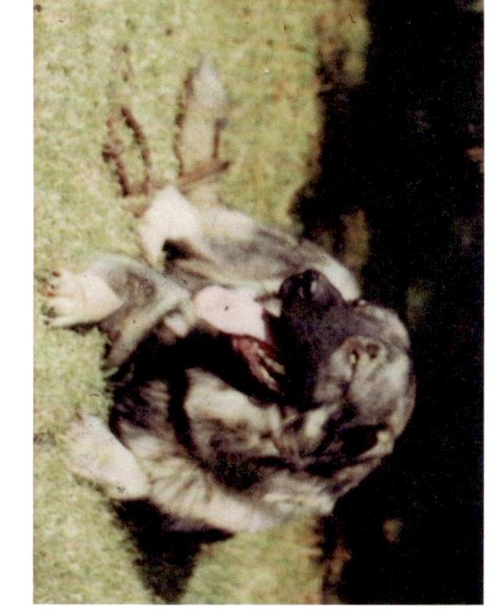

Fig. 58 Mak, Dobova, Slovenia, 1991. © Maja Smrekar

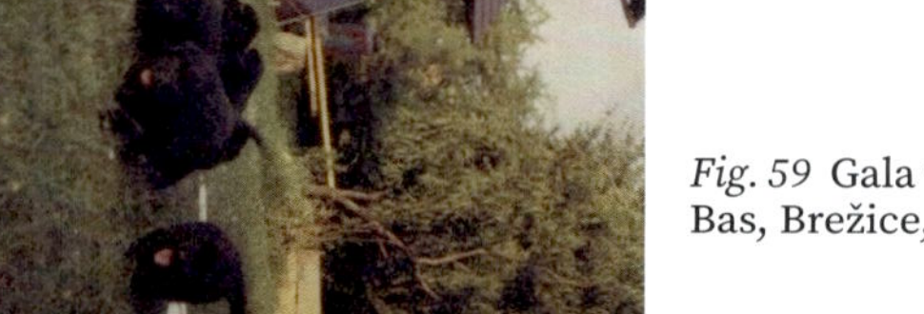

Fig. 59 Gala with the litter she had with Bas, Brežice, Slovenia, 1986. © Unknown

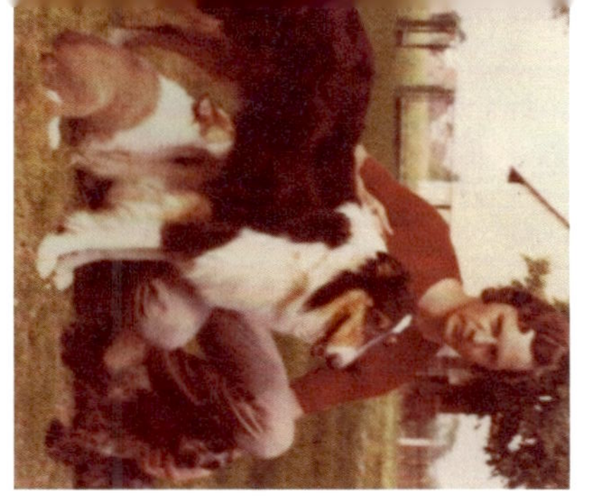
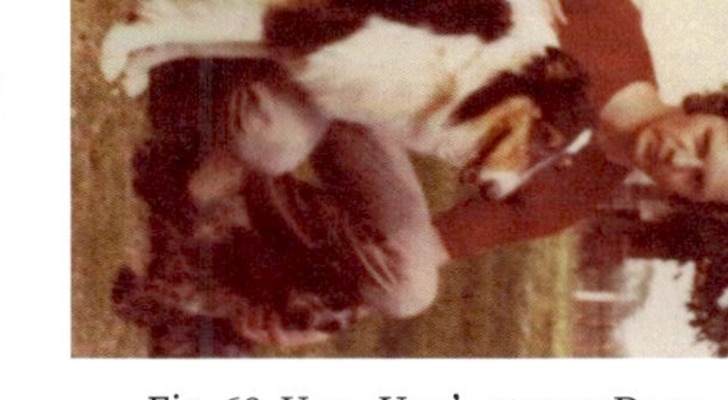

Fig. 60 Una, Una's puppy Doca, and Branko Smrekar in Branko's backyard, Dobova, Slovenia, 1977. © Unknown

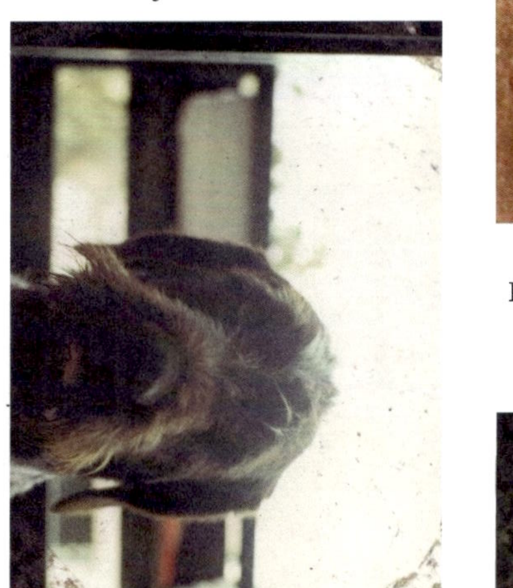

Fig. 61 Ero, Dobova, Slovenia, 1992. © Maja Smrekar

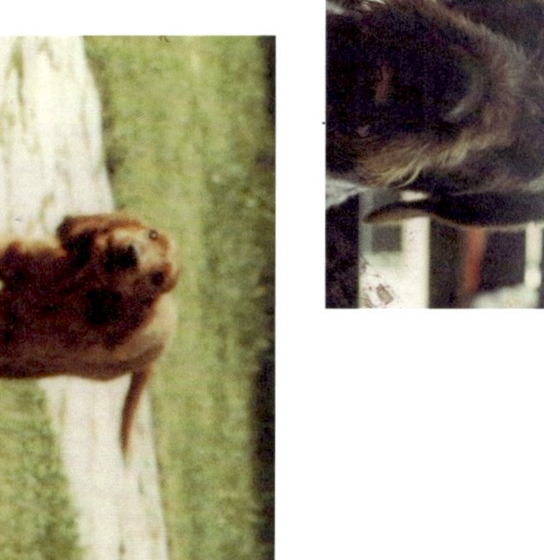

Fig. 62 Rot, Dobova, Slovenia, 1990. © Maja Smrekar

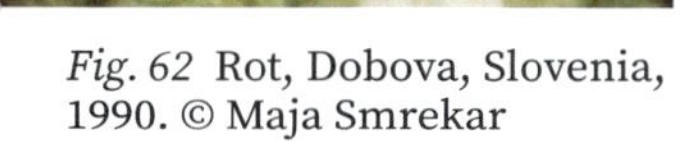

> anthropological machine discourse of dividing species. Our relationship is kinship. When they told me to depict my family in kindergarten, I drew you as my beloved brother. I did not erase you, even though I was five, and my teacher told me that a dog is not and cannot ever be a part of the human family! Nature-culture ends up becoming one word. I want you to make me understand what the difference is between a child and a domesticated animal. Am I a dog? I want to be your dog.

I hope you view my confession alongside my ambition as a point of departure for our dialogue!

Sincerely,
Maja

Copenhagen, November 20, 2015
HYBRID FAMILY

Dear Maja,
I sincerely feel thrilled to be corresponding with you throughout the duration of your overall challenging *HYBRID FAMILY* project, and I am very much looking forward to visiting you and your currently expanding family soon in your Berlin trans-species performativity nest. I hope to be able to accompany your multilayered process of *becoming* and the associated emotional, physiological, psychological, biomedical, biotechnological, bio-political, sociological, and ethical transformations throughout this period, during which you are adopting, nursing, breastfeeding, and raising "your" puppy. When does he-she-it arrive, and have you already "agreed" on a name—and if so, with whom have you agreed?

Your whole *K-9_topology* philosophically and physiologically materializes and condenses such a wide range of both historical and contemporary concerns regarding the role that the construction of otherness plays in the oscillation between selfhood and kinship, borders and membranes, majorities and minorities, "us" and "them."

It seems to stage and carnally enact the collapse of the very dialectic potential of binary wor/l/d-making as such, while, as you rightly state, "Nature-culture ends up becoming one word." Its hyphenation disappears in its outspoken action to become pregnant with meaning, the hyphen itself melting down into an ontologically grey zone of what Scottish anthropologist Victor W. Turner has described as "betwixt and between," as *liminality*.[2]

You, surrogate mother *in spe*, already in the company of your Scottish Border Collie Lord Byron (did you chose him because of his Anglo-Scottish border "origin"?), are obviously not pregnant. You are therefore employing techniques to stimulate breast milk production beyond existing, natural, and supposedly gendered limitations, also implying hormonal changes that are prone to triggering shifts in emotions and empathy. But how do you envision relating your inherent microperformativity of hormones and fluids to ecological and social macro-politics? Beyond narcissism, is "the dog" becoming a mirror? And, beyond narratives of our *bios*, how much *zoe* is there in a dog?

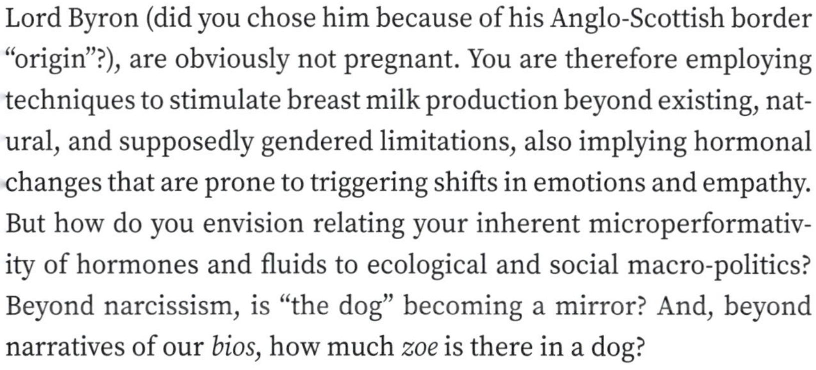

Fig. 63 Police dog Diesel, image published on Twitter by the French Police Nationale on November 18, 2015.

Since you yourself initiated the dialogue with a personal trauma, I cannot help but decide to start with the sinister episode of this week's Islamist terror attacks in my very neighborhood in Paris—where my own (extended, hybrid, but indeed human) family luckily just survived safely. When French police started their anti-terror raid in the suburb of Saint-Denis to capture the alleged mastermind behind the extremely violent shootings, a seven-year-old Malinois Belgian Shepherd police dog named Diesel was killed by the terrorists, in addition to the 130 human deaths that have been deplored so far. According to the French Police, the she-dog was killed precisely when a female suicide bomber activated her explosive vest. Photos of Diesel looking into the camera, side by side with many anonymous uniformed legs and boots, were tweeted, and the hashtag

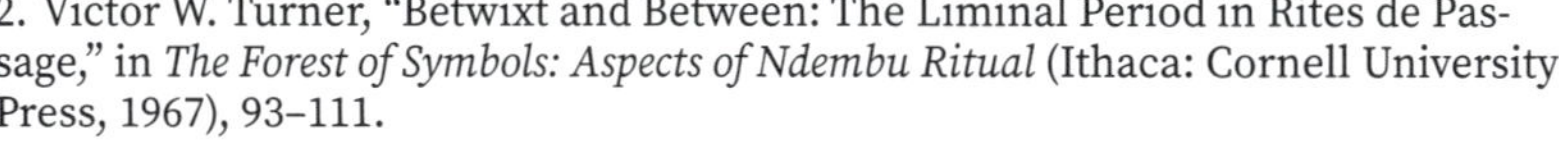

2. Victor W. Turner, "Betwixt and Between: The Liminal Period in Rites de Passage," in *The Forest of Symbols: Aspects of Ndembu Ritual* (Ithaca: Cornell University Press, 1967), 93–111.

Je suis chien, following *Je suis Charlie* deriving from the earlier attacks in January, showcased many "patriotic" dogs bearing French flags and standing on their hind paws. Whatever the final version of the police report says, how did the double feminization—"Diesel against Hasna Aït Boulahcen," the alleged suicide bomber—impact on the commentators' sympathy? Does Diesel also help us to cope with our horror of now seeing even female kamikaze bombers possibly striking at the very heart of our societies against "Western" values, while "Westerners" tend to claim that they are keeping women free from "Islamic patriarchy"? Many commentators sympathize with Diesel-the-dog, ennobled to accompany the list of victims, condemning Hasna-the-human, while the (male) terrorists themselves do not appear on the list at all. Which ingredients are needed to produce victims/heroes, be they dog or human?

In this context, a significant paragraph in Deleuze and Guattari's "Becoming Animal," which obviously served as inspiration for your *K-9_topology,* comes to mind. Instead of simply spotting animal "packs" formed by "contagion," we have a tendency to look for *the* outstanding representative of this minor otherness: "Wherever there is multiplicity, you will also find an exceptional individual, and it is with that individual that an alliance must be made in order to *become-animal.*" In general animality, "there is a leader of the pack, a master of the pack, or else the old deposed head of the pack now living alone"[3]—and why would this not be an exceptionally talented police dog tasked with protecting humans from other human othernesses? Or would Diesel then already have changed sides, despite her ontologically assumed dog-gi-ness? Very often, the act of naming becomes the sign of attributing and recognizing agency ... Moby Dick, Diesel ... Does such naming, then, not only concern those whom Deleuze and Guattari call "Oedipal animals"[4]—individuated animals that can "become" *my* cat or *my* dog—but more generally imply empowerment? Instead of *becoming-minor,* aren't named

3. Gilles Deleuze and Félix Guattari, *A Thousand Plateaus: Capitalism and Schizophrenia,* trans. Brian Massumi (Minneapolis: University of Minnesota Press, 1987), 243; originally published as *Mille Plateaux,* volume 2 of *Capitalisme et Schizophrénie* (Paris: Les Éditions de Minuit, 1980).

4. Ibid., 240.

animals becoming *major* through the very act of naming? Maja, what will be the name of *your* dog? Will it be a *bios* or a *zoe* puppy? I am referring here to a quote from Rosi Braidotti's feminist extension of the concept of Deleuze and Guattari's "Becoming Animal":

> The structural link between women, "native others" and animals has a dense and complex unity; women and "others" personify the animal-human continuity, while men embody its discontinuity. In my language, the former are structurally closer to *zoe*, men to *bios*. The structural link between woman and *zoe* is also a matter of sharing a second-class status, as shown by the relative marginalization of animal life (*zoe*) in relation to discursive life (*bios*). Evolutionary theory supports this by attributing human development to white male skills, while women and "others" are considered mere objects of exchange. Similarly, motherhood has traditionally been considered as an automatic biological process, while fatherhood is seen as a social and cultural institution that rules over and governs biological relations.[5]

In this sense, your *K-9_topology* project in general, and *HYBRID FAMILY* in particular, adds layers to an already rich history of artistic projects in which cohabitation and/or collaboration between human and nonhuman animals—nameless or named—has been supposed to stage and increase empathy. Among all these Kuliks, Albas, Jofis, and Little Johns, we will certainly have a closer look together at how you play with all these imposing references art history has already offered, while you hunt nature and culture hunts you.

Physiologically yours,
Jens

---

5. Rosi Braidotti, *Transpositions: On Nomadic Ethics* (Cambridge: Polity Press. 2006), 104.

Berlin, November 25, 2015
Involution of mOther

Dear Jens,
The last weeks have been filled with research and logistics, accompanied by everyday pathologies, all in combination with my new routine of breast-pumping every three hours for twenty minutes, including during the night, supplemented by mood swings, which are starting to appear on a daily basis, while now following a special diet and hydration protocol.

But I would like to start answering your questions about the name I picked for the now almost eight-week-old baby dog I am going to introduce into my pack as a new family member this Sunday! My choice has been influenced, however, by the meaning of the name of my current dog companion, Byron, a Blue Merle Scottish Border Collie. I therefore have to think back and recall when I first saw the then similarly eight-week-old Byron sitting in a distant corner of the kennel, mischievously observing me—I instantly knew he was the one! His typical, hypnotizing Border Collie gaze triggered my emotional memory and the very first feeling of empathy I developed with a nonhuman: my Newfoundland dog companion Bas, who lived with my family when I was between five and ten years old. I immediately formed an association with a poem, "Epitaph to a Dog," that Lord George Gordon Byron (1788–1824) once dedicated to his dead Newfoundland dog Boatswain. This is how Byron, some years ago, got his name. In order to create a filiation, I therefore decided to name the now newly adopted Icelandic dog by making an associative connection to Lord G. G. Byron's daughter. Due or thanks to her mother's resentment toward her irresponsible, ever-absent father, she was systematically trained in mathematics when she was still just a child and later became the first programmer. In addition, there was another coincidence while deciding on the puppy's name, since all the puppies in the breeder's litter had to be given names starting with the letter L on account of the purebred dog alphabetical naming convention. Lady Lovelace is

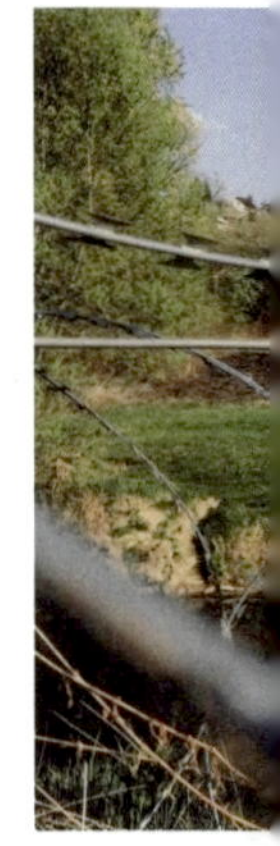

therefore her official name! Byron and Lovelace are thus my nonhuman reminders of art-science synergy as well.

In order to further evoke an archetypical animal, I am actually adding a third L-name to Lovelace: Laelaps—the name of Europa's mythical female dog, referring to the Hellenistic mythology of Zeus and a young Phoenician girl, Europa. She was seduced by Zeus, transformed into a bull, abducted, and carried to Crete on his back. After revealing his true identity, Zeus gave Europa many gifts, among them Talos, a javelin that never missed, and Laelaps, a female dog who never failed to catch what she was hunting. But later on, when Laelaps started hunting the Teumessian fox, one that could never be caught, a paradox ensued: a dog who always caught its prey versus a fox that could never be caught! Laelaps and the Teumessian fox are the embodiment of what is called the irresistible force paradox: "What happens when an unstoppable force meets an immovable object?"

I therefore see the current European refugee crisis, which coincides with the catastrophe of the Paris attacks, as part of the same paradox: the EU is fighting terrorism even though the EU is the one that created it in the first place. Maybe grieving for Diesel was a convenient gesture instead of, for example, investigating why the Paris terrorists were actually French and Belgian citizens. How can it be that socially non-integrated young people in Europe decide to carry out such a horrific, radical act? Despite all the empathy we have with all the people traumatized by the Paris attacks, can we imagine just for a moment that, maybe for Hasna, her gesture was more an act of class war than of feminism?

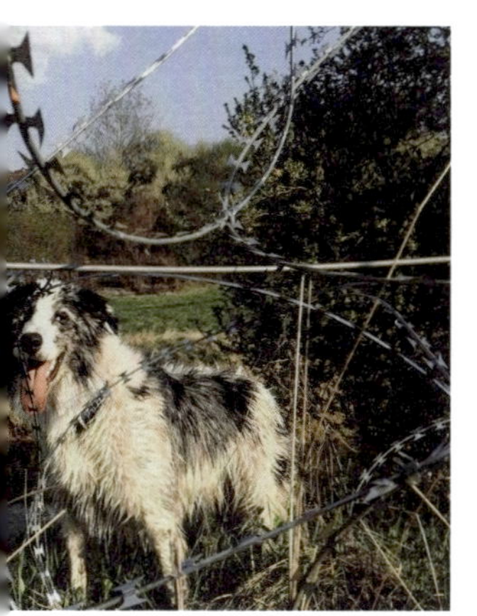

Fig. 64 Byron at the Schengen Green Border between Slovenia and Croatia, 500 meters from Smrekar's family home, Dobova, Slovenia, 2016. © Maja Smrekar

But then, how can we artistically address larger issues such as the European crisis through a molecular perspective? I feel that I need to start mounting resistance toward this crisis on a personal level. I first need to exercise power on and through my body in a different way, and likewise allow for my dog companions to regain theirs. Your remark, Jens, that I am "pregnant with meaning," instead of being pregnant physiologically, has been inspirational for me. By being pregnant with a meaning, I am becoming mOther! This process therefore involves the intimate, the social, and the political at the same time.

By breastfeeding LLL / a dog / a nonhuman (?), my aim is to resist the conservative neoliberal system of biopolitics that rules the whole world as we know it, down to our smallest subatomic particles. *Becoming* is a molecular process: in my case the molecular process of my pituitary glands being stimulated by systematic breast-pumping in order to produce the hormone prolactin and ultimately accumulate milk. One side effect of this is that the level of the hormone oxytocin increases, favoring empathy and contributing to the phenomenon of "trans-bordering." On the other, biopolitical side, we are ruled by borders as a political concept and therefore exert power over others, with or without consent. I grew up on the Slovenian-Croatian "Schengen" border … . Maybe this made me—unconsciously—choose the Scottish Border Collie Byron as a canine nomad companion, with whom I have been traveling all over Europe, across all the boundaries and their ensuing laws and rules.

Last weekend, I again witnessed and crossed borders, through my oxytocin-haze-filtered perception, driving a car over barbed-wired, heavily armed, traumatized Europe on my way from Berlin, via Belgium to Paris for a conference and back. All the borderless transitions between the above-mentioned countries turned into heavily-armed check points. Sometime in the middle of the night, my MP3 player started playing Mozart's *Lacrimosa* while I slowly drove in the long line of cars. At the border to France, armed policemen pointed a strong, huge, and blinding light cannon straight into my face. In that very moment, I received yet more confirmation of my molecular performativity process, increasing the amount of oxytocin in my body. That was the moment I made my decision: empathy is my form of resistance to the cynicism of Europe!

Sincerely yours,
Maja

Los Angeles/Mexico DF, December 9, 2015
HYBRID FAMILY—Involution of mOther

Dear Maja,
Your oxytocin blast while practicing breast-pumping not only seems to have strongly impacted on your last writings, but has rendered me—although digitally abstract—globally emphatic when traveling across different biological and cultural spheres, currently on the other side of the Atlantic. While reading your account of the mythology and etymology of canine cultural references, I suddenly see all relationships and behaviors through the filter of your *K-9 topologies*: whether my observations of and interactions with a typical miniature lapdog belonging to an over-caring, childless Californian couple whose universe entirely comprises the interiors of the home, the car, and the pet hotel, or with the disregarded, stray under-dog in Mexico City while paying attention to the facial expressions of mystical breeds such as the chihuahua, the smallest recognized breed, or the sensitive Mexican Xoloitzcuintli hairless dog—I feel tempted to detect mutual domestication and hybrid projections everywhere.

Reading, then, about your *INVOLUTION OF (m)OTHER*—a brilliant pun, chapeau—your embodied paradox becomes even clearer, since the term "involution" usually refers to a decrease in the size of an organ, usually associated with a decline in function, such as the uterus, the ovaries after menopause, or the milk-producing lobular structures of the breast. It struck me when thinking and feeling that you are, right at this moment, getting ready, psychologically and physiologically, for the great day when "your" selected adoptive puppy will arrive, and you'll check whether she is even as ready to engage in this trans-species mOtherhood as you are: despite the hard corpo-real efforts that your diet and the constant, regular pumping at these exhausting three-hour intervals must imply, and the excitement I imagine you are feeling, you are focusing a lot in your writing on the underlying significance of giving the "right" symbolic name to the creature arriving into your "pack" and onto your breast. Involution of the uterus without pregnancy, a mOTHER pregnant with meaning? Semantic hyper-compensation while awaiting the visceral?

I am trying to understand your emphasis on *involution,* less in the aforementioned medical sense and more as a way to oppose the notion of *evolution,* just as Deleuze and Guattari resisted the logic of *filiation* by replacing it with *contagion,* according to the first principle of *becoming-animal:* "pack and contagion, the contagion of the pack, such is the path becoming-animal takes."[6] Contagion of the pack through *adoption,* which factually it is, just like the much-quoted "*aparallel evolution,*"[7] with which Deleuze and Guattari also praised viruses in particular as the crucial and active vectors responsible for liŷe's diversity. A virus can "move into the cells of an entirely different species, but not without bringing with it 'genetic information' from the first host. ... Transversal communications between different lines scramble the genealogical trees. Always look for the molecular, or even sub-molecular, particle with which we are allied."[8] But, beyond hermeneutics and beneath pharmaceutics, what does the whole poly-*sensual*-beyond-the-poly-*semic* feel like?

The reason for my question is the following: We know that in the history of art, creative impulses have often been imputed to forces of otherness external to the artist "genius," forces that supposedly escape cognitive control, so as to make him (most often) or her (less often) the very medium through which mysterious, universal, or natural forces materialize and manifest themselves as the artist's art. Articulated in formulas such as Arthur Rimbaud's famous "Je est un autre,"[9] the homogeneous, conscious subject is challenged by manifold drives—but most often by spiritual forces, psychological splits, not by physi(ologi)cally explainable alternatives to human identity, selfhood, or agency based on the natural sciences. And while the recourse to animality and related allegories as a vector of representation in art is a common feature, the challenge of your *HYBRID FAMILY* project is precisely its radical embodiment and re-materialization of symbolic layers that have been compiled over

6. Gilles Deleuze and Félix Guattari, *A Thousand Plateaus,* 243.

7. Ibid., 10.

8. Ibid., 11.

9. Arthur Rimbaud, Lettre du Voyant, to Paul Demeny, May 15, 1871, in *Correspondence inédite (1870–1875)* d'Arthur Rimbaud, ed. Roger Gilbert-Lecomte (Éditions des Cahiers Libres, 1929), 51.

and over again in our Western culture of interpretation—which lacks these *presence effects*, to speak with Hans Ulrich Gumbrecht.[10]

Meanwhile, the puppy, she, Lady Lovelace Lealaps, must now have arrived at your post-Beuysian Berlin Bülowstraße loft! I am curious to viscerally apprehend how you like Lady Lovelace Lealaps and how Lady Lovelace Lealaps likes you. How do your microbiome communities match when, as Donna J. Haraway states, you are "in company with these tiny messmates—to be one is always to *become with* many"?[11] How does she smell? How do you smell to her? Can you grasp a certain biosemiotic level of molecular, olfactory exchange? Does she effectively suckle you as you had hoped? Does she like your milk? And also: Did you ask Lady Lovelace Lealaps if she agreed to be adopted? And—how—did she respond?

The question of animal consent I am posing here is certainly not new; it has also been addressed by "practical ethics" philosopher Peter Singer with regard to zoophilia, which, according to him, is an acceptable practice as long as there is mutual consent. But the question also popped up when curating the trans-species blood brother— or sister—hood performance *Que le cheval vive en moi* (May the Horse Live in Me) by French artist duo Art Orienté Objet in Ljubljana in 2011: artist Marion Laval-Jeantet turned herself into a proverbial guinea pig, allowing herself to be injected with horse blood plasma containing the entire spectrum of foreign immunoglobulins, glycoproteins that circulate in the blood serum. But she would not participate in a transfusion to the actual animal, because she was unable to ascertain whether the horse agreed, or how to *respond*. In their famous, "viral" internet music video *What Does the Fox Say*, Norwegian comedy duo Ylvis mischievously asks, "If you meet a friendly horse, will you communicate by mo-o-o-o-orse?"[12]

One of the most inspirational philosophical essays about animality is certainly Jacques Derrida's *L'Animal que donc je suis (à suivre)*

10. Hans Ulrich Gumbrecht, *Production of Presence: What Meaning Cannot Convey* (Stanford: Stanford University Press, 2004).

11. Donna J. Haraway, *When Species Meet* (Minneapolis: University of Minnesota Press, 2008), 4, original emphasis.

12. "The Fox," *azlyrics*, accessed July 11, 2024, www.azlyrics.com/lyrics/ylvis/thefox.html.

[*The Animal That Therefore I Am (More to Follow)*],[13] in which the questions both of calling animals "animals"—and giving them names, or not—and of how so-called animals may respond become central. "Animal is a word that men have given themselves the right to give. These humans are found giving it to themselves, this word, but as if they had received it as an inheritance."[14] Naming—really contamination, or filiation, then? And further: "All the philosophers ... say the same thing: the animal is without language. Or more precisely unable to respond, to respond with a response that could be precisely and rigorously distinguished from a reaction, the animal is without the right and power to 'respond' and hence without many other things that would be the property of man."[15] What does it mean to name a living being that is—not unusual for dogs—eight weeks old, still unnamed, and getting a new "owner," a person entitled to say "my dog"? *Nomen est omen*, you write, but for whom? More to follow ...

I follow you, indeed, in your desire to deconstruct what you have often addressed in our previous discussions as "the fascist system of dog breeding," with its obsessions with pedigree and restrictive nomenclature when it comes to the mandatory letters—L—that must be used when giving names to the offspring from a single litter. Lovelace Lealaps sounds like a cynical and resolutely post-Linnaean (another L) nomenclature. And the feminist evocation of British mathematician and writer Ada Lovelace—let's appreciate the A, then, from now on—in the context of a nonlinear pack, with her reputation of having conceived the first algorithm said to have been performed by a calculating machine, appears to me to be reminiscent of when Donna J. Haraway, in *When Species Meet*, illustrates her writings with a cartoon by Dan Piraro in which the "American Association of Lapdogs" gathers in order to contrive a strategy to combat their worst enemy—the laptop![16] Lovelace against laptop, what a paradox.

13. Jacques Derrida, "The Animal That Therefore I Am (More to Follow)," trans. David Wills, *Critical Inquiry* 28, no. 2 (Winter 2002): 369–418; originally published as "L'Animal que donc je suis (à suivre)," in *L'animal autobiographique: Autour de Jacques Derrida*, Actes du colloque tenu à Cerisy-la-Salle 11.–21.7.1997 (Paris: Éditions Galilée, 1999).

14. Ibid., 400.

15. Ibid.

16. Haraway, *When Species Meet*, 9.

Your mythological Lealaps dog that always makes her catch, in paradoxical competition with the Teumessian fox, who always manages to escape, points more generally to the role of animal allegories in human mythology, and often in paradoxes. Think of Zeno of Elea's paradoxes, such as the one about Achilles who, though reputedly fleet-footed, would never have been able to overtake a tortoise after generously giving it a few meters head-start—something disproved by the simplest of real-liye experiments. With regard to such ancient sophistry, we may find two types of epistemological reaction: on the one hand, that of logicians, starting out from the actual logic of known elements and demanding that one distinction be drawn between levels of language and another between the subject of the utterance and the subject of the statement; and, on the other hand, a more pragmatic, common-sense attitude, like that attributed to Diogenes the *Cynic*, who, on being presented with a paradoxical argument against the existence of movement, made no answer and ... just walked away. Maja, you reminded me that the term "cynic" actually derives from the Ancient Greek *kynikos*, dog-like, since the cynics were insulted "as dogs" because they proudly rejected conventions to provoke alternative thought.

Do your Scottish Border Collie Lord Byron and Lady Lovelace Lealaps become "non-human symbols"? *What does the dog say*? In the tradition of biosemiotics, the concept of *interpretation* is not a human exclusivity; here, all biological systems employ sign relationships that are constantly being interpreted. The symbolic, however, seems to remain a sign of high semiotic complexity. As biosemiotician Jesper Hoffmeyer writes, "... humans possess the ability to communicate and think via symbolic references, while all other organisms seem to be limited to iconic and indexical referencing."[17] In this sense, I'm curious about how, going beyond the symbolic naming procedure, your other senses are now becoming engaged in your bodily relationship of hybrid mOtherhood.

Yours,
Jens

17. Jesper Hoffmeyer, "God and the World of Signs: Semiotics and the Emergence of Life," in *Zygon* 45, no. 2 (June 2010): 372.

Berlin, January 3, 2016
Birth of the Kennel

Dear Jens,
Today I want to share some thoughts on very different levels—associative thinking is currently the only modality I am capable of at this stage of the *HYBRID FAMILY* process, entangled with strong emotions and overwhelming hormonal changes. I am therefore offering a palette of bits and pieces in what might resemble a sort of Proustian croquis style.

The 29th of November was already the day to pick up—Ada—in the picturesque Schwarzwald village of Haigerloch. The night before, I was woken by a panic attack and wasn't feeling well on my way to the village, driving my old car from Berlin down to the Schwarzwald and back—almost twenty hours including roadblocks and storms. Every three hours, I had to stop, install the cyborg-like breast-pumping equipment under a huge pullover, and drive further for the next twenty minutes, and then stop again to remove the equipment. I was feeling both horny and depressed. Horny, because the oxytocin kicks in all over my body while breast-pumping. Depressed, because I've been feeling like an ultimately vulnerable underdog. Byron, my actual companion dog, stoically accompanied his childless human fellow on our shared odyssey marked by my anxiety attacks ... and pleasure ... every three hours for twenty minutes. All the rest was just fear. I started wondering, though, whether I am afraid to fail as an artist or as a mother?

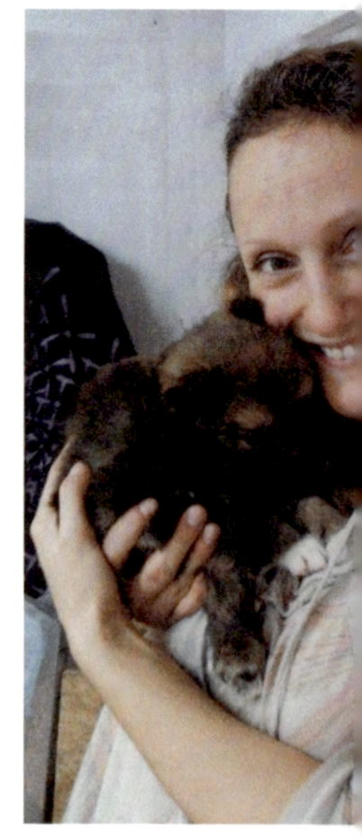

I believe that my culturally coded anxiety connects me with one of Rosi Braidotti's thoughts you shared with me earlier: " ... motherhood has been considered as an automatic biological process, while fatherhood is seen as a social and cultural institution that rules

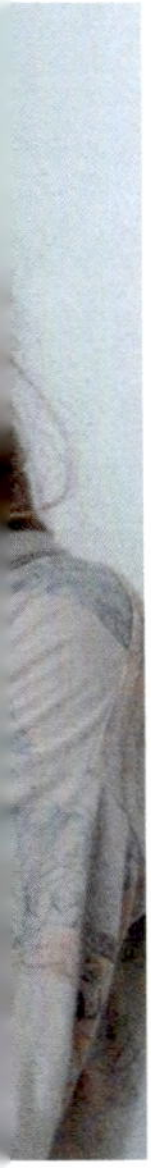

Fig. 65 Smrekar meeting one-month-old Ada for the first time, Haigerloch, Germany, 2015. © Katarina Hergouth

over and governs biological relations." I am transforming into a creature of social reality as well as a creature of fiction by multi-directionally exchanging the flow of bodies, fluids, and values—with a dog! It's scary yet also poetic. Zoopoetic. I find a temporary shelter in the words of Carolee Schneemann: "My advice to younger artists: give yourself permission."[18] Indeed a great piece of advice from a good mother! Still, it is hard to remain powerful, dignified and ... above all, in control.

Anxiety aside: physiologically executing my process within the discourse of a molecular performance and hormonal biopolitics via my collaboration with Ada, constantly repeating the act of breast-pumping, and hence training my pituitary glands, I am becoming aware that the training procedure with my own body totally resembles that of dogs being clicker-trained. I have been carrying out the same procedure thanks to my amygdala. Or is my amygdala training me, by letting me know when it is time to start experiencing these somatic sensations?

On that note, I do really think we should talk through "Morse with a horse," as you suggested, since domesticated animals as well as their human counterparts are somehow "bilingual." We inter-communicate via a language that is neither human nor animal, but hybrid. Dogs communicate their willingness or unwillingness to collaborate very directly at any moment, especially when they face states of fear, anxiety, or even agony.

Indeed, great news! After only pumping out some colostrum during the whole month of December, I finally got some drops of real milk—the first one just a day before Christmas. What an irony!!! After waiting patiently, sitting on the floor with Byron, Ada finally approached me, smelled the milk and licked those few drops as they ran down my belly. She—logically—traced the drop to its source, which she then repeated every time she heard the pump stopping. It all happened smoothly and organically. This is how a-linguistic unmeaning happens.

18. Talitha Kotzé, "Iconic America Artist Carolee Schneemann Appearing at the Edinburgh Art Festival 2012," *Summerhall*, July 31, 2012, accessed July 11, 2024, www.summerhall.co.uk/press/iconic-america-artist-carolee-schneemann-appearing-at-the-edinburgh-art-festival-2012/.

The whole month of December in Berlin was challenging, including the holiday atmosphere, in which consumerism and religious institutions hold hands. Now, at the beginning of January, I am still not getting drops of milk every time I pump, even though it's been happening on a more regular basis. Sometimes I don't get a drop of milk for a few days. Speaking with Donna J. Haraway, it is love that functions as an ethical bind between the companion species: love here names the ethical co-emergence and cohabitation between specific, historically situated dogs and humans.[19] Love therefore does not mean a mere intensification of affect, or a more proprietary form of possession ("I love animals"), but rather a preparedness to examine the interrelations between humans and other species. As Charles Darwin wrote in his 1838 notebook, "He who understands baboon would do more towards metaphysics than Locke."[20]

Byron has now been helping me a great deal with raising Ada. Whenever Ada does something she should not, he arrives immediately and growls at her until she refocuses. He is helping me to set boundaries for her. We are therefore parenting Ada together. The three of us hug and kiss, lie in bed, crawl on the floor, jump on and off the furniture, scream and growl within our process of getting to know each other, (re)setting our boundaries. There are three of us, as there were in my former, human, biological family. However, our dynamics are very different. We are a pack in which I am the "alpha"—as the provider of food and shelter, then Byron as the experienced companion, and then baby Ada. Nevertheless, they are the ones teaching me how to set boundaries, a skill which my parents failed to teach me.

I guess my performance in the Bülowstrasse residency apartment might bear some similitude to that of Joseph Beuys and the coyote Little John in their shared gallery space in New York. Neither the human nor the animal was at home in this space, but by engaging with various props, the two participants built a sensitive agreement

19. Donna J. Haraway: *The Companion Species Manifesto: Dogs, People, and Significant Otherness* (Chicago: Prickly Paradigm Press, 2003).

20. Charles Darwin, 1938 notebook entry, quoted in Dorothy L. Cheney and Robert M. Seyfarth, *Baboon Metaphysics: The Evolution of a Social Mind* (Chicago: University of Chicago Press, 2007), 4.

about a shared world. The performance of fashioning this space shared by the human and nonhuman agents was itself crucial to the artwork—much like the nursery room I've been building for the final *HYBRID FAMILY* performance. My temporary home, this performance space, exists between inclusion and exclusion, like a camp, holding a status of exception.

I am very much looking forward to you visiting our *HYBRID FAMILY* camp in a few weeks' time. I hope I'll be feeling better by then—otherwise you can expect a palette of lateral verbal associations, accompanied by growling, biting, dropping, dripping, and many other forms of nonverbal communication.

A big hug,
Maja

Paris, January 22, 2016
HYBRID FAMILY—Birth of the Kennel

Dear Maja,
Following your advice, I have been reading your report in the manner of a Proustian croquis, from the epic journey to pick up Lady Lovelace Lealaps, alias Ada, and your co-parenthood with Lord Byron, to the first swallowed drops of your childless mother's milk after all those horny-depressive breast-pumping efforts. I have been trying to oscillate between, on the one hand, my own deep emotional and co-corpo-real involvement and, on the other, a more distant and somewhat abstract view of your process, not unlike a biosemiotically motivated entomologist.

I can't wait to soon go beyond the level of our epistolary exchange and visit you and the hybrid family when you open up your Bülowstrasse space—next week! The format of epistolary exchange we chose ultimately has not resulted in a digitally flattened acceleration of quick updates and anecdotes, but rather in somehow parallel and "autistic" essays, each with different rhythms and their own styles, quotes, associations, and seemingly unrelated references at first sight, to fill up the imaginary with complementary elements not foreseen in the protocol.

You, self-diagnosed alpha-female, had indeed warned me at the very beginning of the project that your process of becoming mOther might unfold as a "narcissistic ego-formation." Counting many performance artists among my appreciated friends, who all engage their bodies and biographies in their work, I have always wondered about the fragile balance between narcissism, altruistic engagement, subjectivity, self-righteousness, and universal claims. There is a transhistorical reason for the everlasting form of the artistic self-portrait and its subsequent contemporary developments, and it may well be that the personal union of the artist and the work unleashes a creative energy dissimilar to the kind that is generated when the artist steps back to take a third-person perspective. At least three different aspects come to my mind when I anticipate talking to you about your *HYBRID FAMILY*, *in situ*, in one week's time:

- One might be intrigued by the way you describe the process of communal parenting, self-domestication, and education becoming entangled. How much self-effacement does it require to transform trans-species communication into a material form of attunement?

- To what extent is such a singular narcissistic experience, as you frame it, indeed synthesizing and collaging the numerous artistic projects that have previously investigated the human-dog-wolf trope, and how do you reference and transform those inspirations?

- How will this multisensory and multimodal process be made culturally apprehensible to visitors sharing this unique experience? Not just as a staged image, as we humans are, indeed, visual animals ...

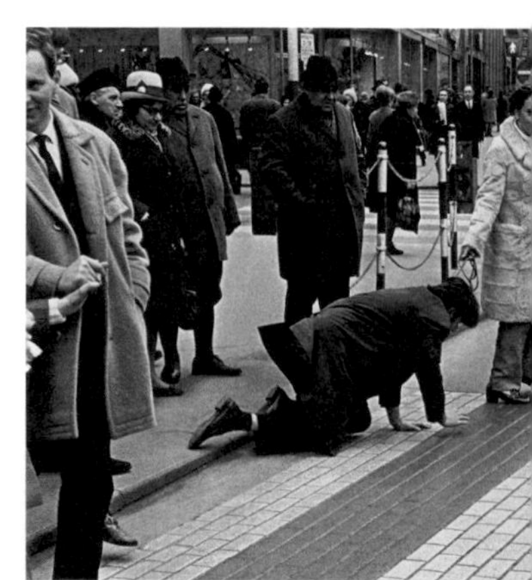

Fig. 66 VALIE EXPORT and Peter Weibel: Aus der Mappe der Hundigkeit (From the Portfolio of Doggedness), 1968, Generali Foundation Collection—Permanent Loan to the Museum der Moderne Salzburg. © Generali Foundation / Bildrecht, Vienna, 1968, photo: Josef Tandl

You will be spending a week in the company not only of Ada and Byron but also of performance photographer Manuel Vason, who has collaborated with more than two hundred performance artists in very intimate and immersive ways. I am extremely curious about which of these kinesthetic and intimate vectors of fluid biopolitics and interactions will be captured. I hope to overcome my skepticism stemming from bad experiences that I have had when performance art has been reduced to gestural codes in isolated snapshots, resulting in the kind of stereotyped Warburgian *Bilderatlas* with which art history students are still being introduced to performance art today. I admit to being equally annoyed when performed actions are just staged for the very purpose of being documented, and I still feel traumatized by a 2004 performance by Marina Abramović and Jan Fabre at the Palais de Tokyo in Paris, *Virgin Warrior*, in which both performers wore armor inside a giant glass box in the middle of an audience. All theatrical gestures were choreographed just for the multi-camera shooting, the precise moments in which Marina's tears ran down her cheeks, etc., but relations with the audience did not matter at all—bad cinema and theater.

Conversely, while your *HYBRID FAMILY* acknowledges manifold previous art projects dealing with the human-dog-wolf trope, your molecular, microperformative, and physiology-centered approach goes a step further, creating a kind of multisensory supercollider for these references. How can I not think of Eduardo Kac's project of a genetically engineered, fluorescent dog, *GFP K-9*, while Kac himself quotes Nam June Paik's collapsing *K-456* robot, cited again in France Cadet's strangely programmed greenish robotic dogs? In your previous performance *I Hunt Nature, and Culture Hunts Me*, you directly referred to Oleg Kulik as a social, political, and artistic "humanimal."

Beuys's interactions with the "wild" coyote contrast with the anthropomorphization of dogs—e.g., in Wim Delvoye's sketch for a surgical operation to give a dog a human face via rhinoplasty, or, at the other end of the scale, William Wegman's staging of rowing dogs wearing liÿe vests. In her feminist critique of gender roles,

VALIE EXPORT walked Peter Weibel as her dog in front of a Humanic shoe shop in *Aus der Mappe der Hundigkeit*. We see a nice correspondence between your performance *Ecce Canis,* which dealt with the similarity of genetic sequences that evolved in parallel in humans and dogs, and Klaus Spiess and Lucie Strecker's *Hour of the Analyst Dog,* in which they reenacted the NC_006592.3 sequence extracted from Sigmund Freud's chow chow.

While such influences and references, and their respective narratives, all seem to resonate in your *HYBRID FAMILY,* its sociopolitical dimension transforms it into a liýetime commitment; meanwhile, the hormonal and molecular transformations imply radically shifting from the metaphorical to the metabolic level. I anticipate that our encounter at your trans-species Bülowstrasse nest next week will likewise be not only an ephemeral rendezvous between an artist and a curator but also the start of an equally long-term commitment and exchange.[21]

Yours,
Jens

Kamenjak, Croatia, July 15, 2016
Manifesto on mOtherness

Dear Jens,
I have something strange to confess: one day last month, I dreamt I was in labor! I was on my way to the hospital, but there were just so many things and personal issues to resolve, so many people to talk to, so that, even when my cervix started opening, I still couldn't manage to get to the hospital. Today, it has not only been precisely one month since that dream but also exactly nine months since I

21. The encounter between Ada, Byron, Smrekar, and Hauser took place on Friday, January 29, 2016, at Freies Museum Berlin, Bülowstrasse 90, 10783 Berlin, Germany. For the whole week, Smrekar, Ada, and Byron welcomed small groups of guests each day for informal sessions of exchange, which included breast-pumping, nursing, and teatime, before they definitively settled in as a family in Ljubljana.

started the *HYBRID FAMILY* process in October 2015. And I woke up knowing the time had come to write to you again.

The *HYBRID FAMILY* has since then been continuing in real life, as Ada moved with me and Byron to Ljubljana and has been taking part in our daily psycho-pathological battle for survival. We've been contributing to the grim spectacle of neoliberal capitalism that levels human and nonhuman animals in their interchangeable place within the global market economy.

Fig. 67 Smrekar and Bas, Dobova, Slovenia, 1984. © Branko Smrekar

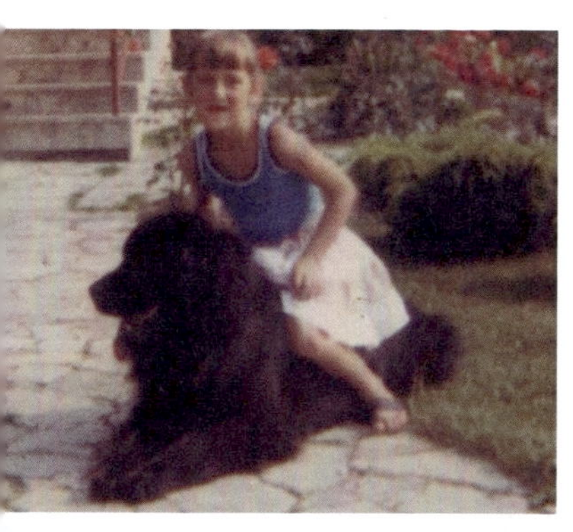

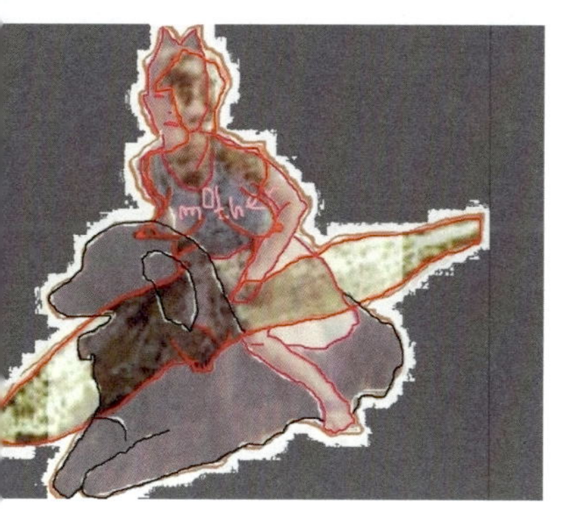

drawn during the preparations for the *K-9_topology: HYBRID FAMILY* performance, Berlin, Germany. © Maja Smrekar

After Manuel Vason's visit, where he lived with us in our private hybrid family seclusion for a week in early January, resulting in a series of truly sensitive photos emerging throughout his stay, I fondly remember you being the first visitor to the *HYBRID FAMILY's* public *becoming*. During the following five days, we always welcomed three people at a time, as a kind of "performance" in that apartment that had been transformed into a nursery room. Ada, my intelligent darling, connected so gracefully with me throughout the whole process. She would run up to me each time she noticed the breast pump stopping in order to suckle some milk from my breast; meanwhile, visitors were drinking galactagogue herbal tea and discussing the various layers of meaning in the transition from a human to a non- or more-than-human family. They would witness not just Ada feeding on my colostrum but also all the profound gestures connecting us two beings. But the whole four months of seclusion became an especially meaningful process to me as well for other reasons linked to the microperformative hormonal stimulation of breast-pumping: I realized that I was performing the same strategy as some mothers-to-be about to adopt a still nursing baby, for which one requires neither ovaries nor uterus. This process of breastfeeding is not necessarily connected to pregnancy and can potentially be carried out no matter

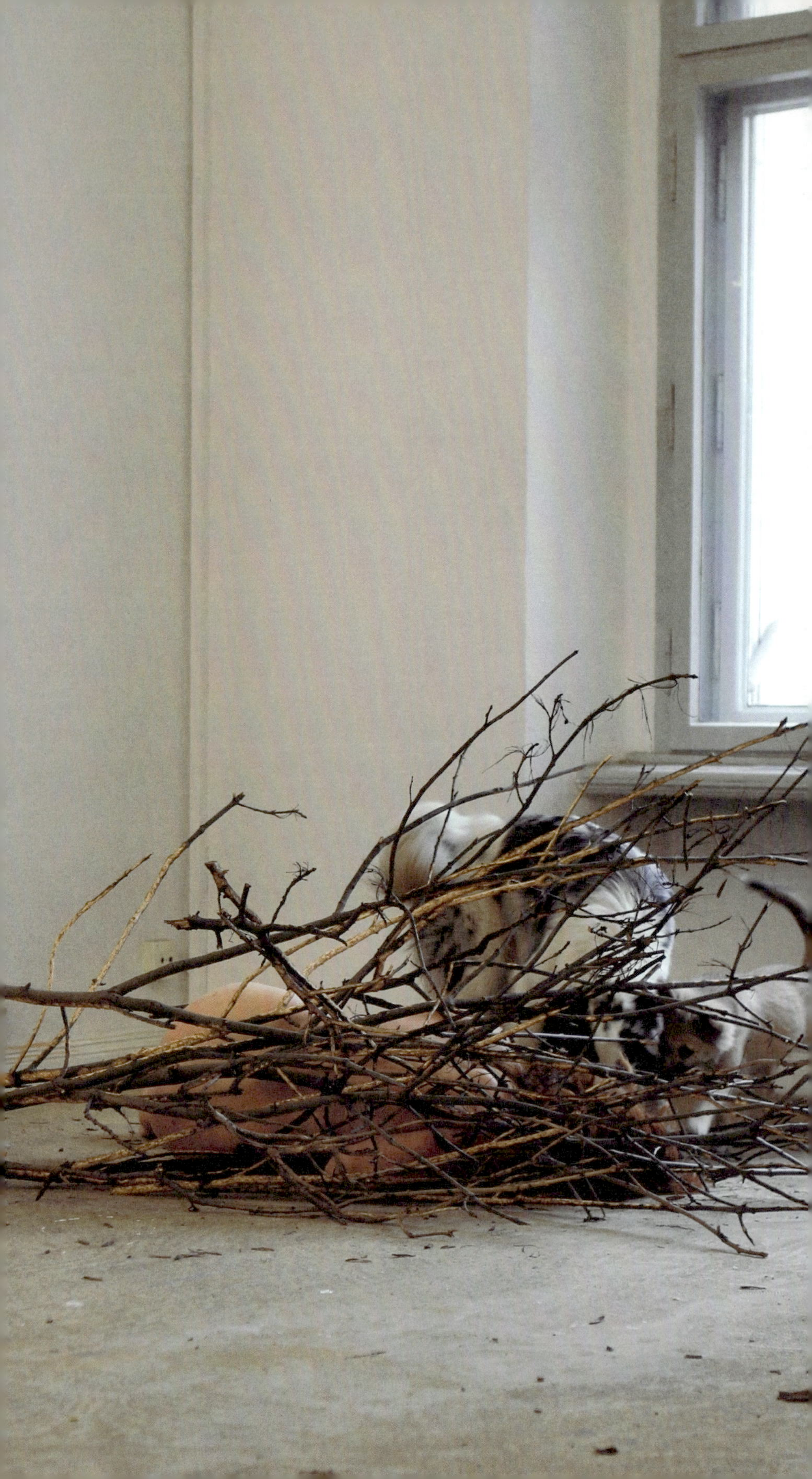

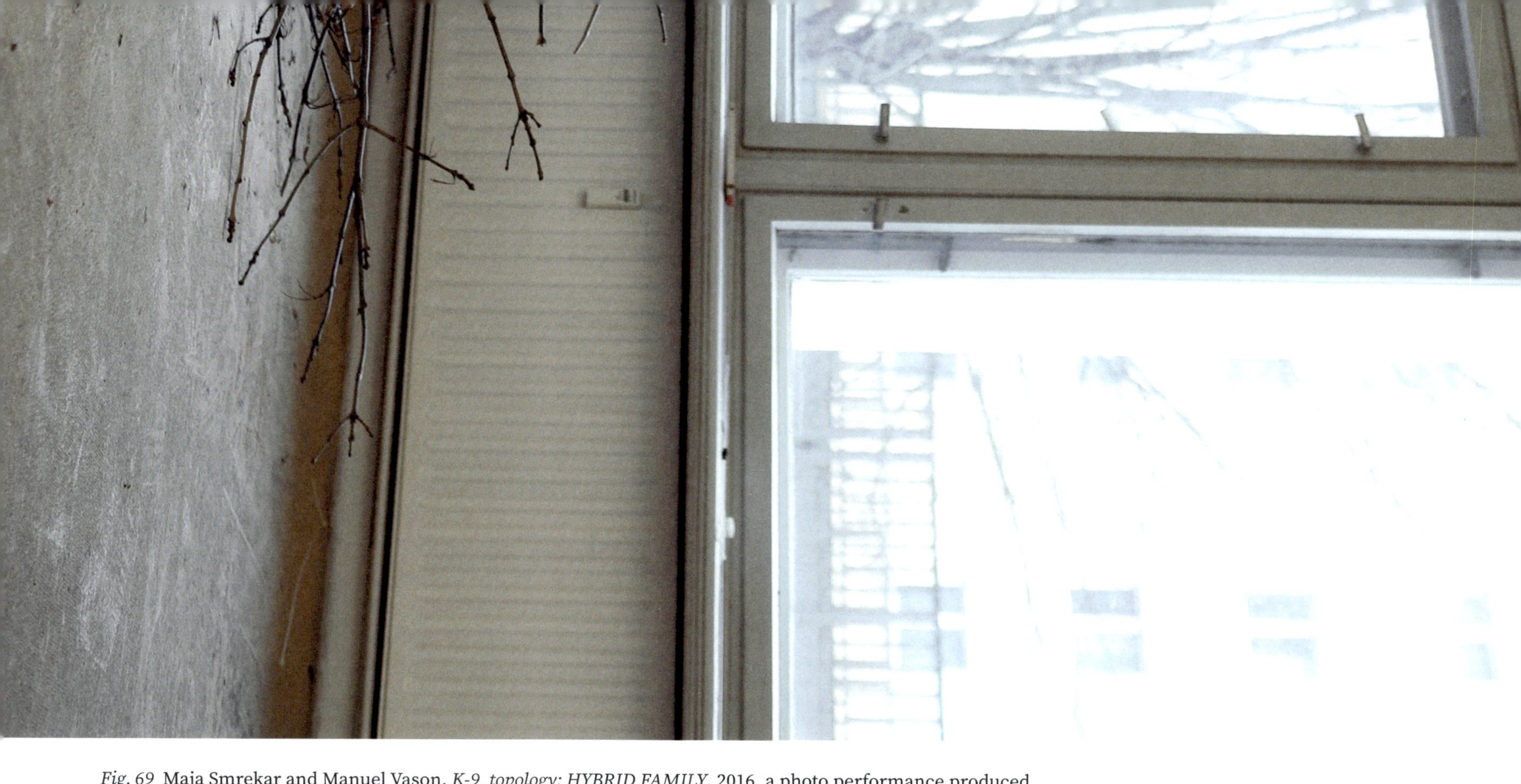

Fig. 69 Maja Smrekar and Manuel Vason, *K-9_topology: HYBRID FAMILY*, 2016, a photo performance produced by Kapelica Gallery, Slovenia, and Freies Museum Berlin, Germany. © Maja Smrekar and Manuel Vason

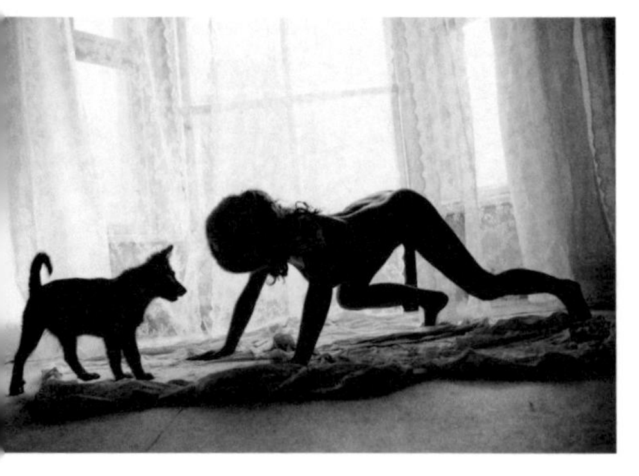

Fig. 70 Maja Smrekar and Manuel Vason, *K-9_topology: HYBRID FAMILY*, 2016, a photo performance produced by Kapelica Gallery, Slovenia, and Freies Museum Berlin, Germany. © Maja Smrekar and Manuel Vason

one's gender or reproductive ability, which I find an important statement in the project.

This hybrid experience also helped me to answer a question that Donna J. Haraway poses: "What is decolonial feminist reproductive freedom in a dangerously troubled multispecies world?"[22] My personal answer might sound as follows: The Anthropocene is about the destruction of places and times, which makes me doubt whether I could become a biological parent to a human baby. Consequently, while I regret not being a mother, I would just as much regret being one. However, just because I don't have biological children, it doesn't mean I do not have maternal instincts. Even though the hormone prolactin ensures the survival of the species via its reproductive function, it also ensures the survival of individuals across many species in its homeostatic functions, as it plays a role in many other metabolic functions in our bodies outside of breastfeeding. My experience is that the microperformative interplay between the two hormones (prolactin and oxytocin) produced in my body by breast-pumping, and later by breastfeeding, increasingly engendered an enhanced feeling of empathy. This empathy confirmed my point of departure for the *HYBRID FAMILY* paradigm: by becoming pregnant with the concept of abundance, I became Other as mOther. I'm convinced that we will see endless debates on this subject in the future. For me, it is clear that motherhood, rather than being reduced to parenthood only, should become abundant within interpersonal relationships at large. It should transform into *mOtherness* as a concept of solidarity, allowing for the cocreation of responsible futures among humans and other-than humans. I therefore call on and welcome (women of) all genders to start and/or (continue) breastfeeding!

22. Donna J. Haraway, *Staying with the Trouble: Making Kin in the Chthulucene* (Durham: Duke University Press, 2016), 6.

This exclusive dialogue with the large variety of insightful thoughts has been a true honor to share with you, Jens. Let's celebrate all of our dearly beloved hybrid Others!

A furry, fuzzy hug from
Ada, Byron, and Maja! ***

Post Scriptum: TO ADA

The *HYBRID FAMILY* project was part of *K-9_topology*—a series of projects that arose out of Smrekar's fascination with dogs. They had been family to her as long as she could remember, which was the reason she invited some canine individuals to become artistic collaborators. Her dog companion Ada was one of those nonhumans who showed a great interest in, and a deep devotion to, their shared artistic processes. Her presence was not only the quintessence of the *HYBRID FAMILY* but of many other artistic projects, such as *ARTE mis* (2017), *Opus et Domus* (2018), and *Brute_force* (2020–22), which she contributed to with her emotional intelligence.

Ada passed away suddenly on May 10, 2024, at the very time that this public dialogue was being edited for this publication. She will not only be dearly missed but will also remain an important figure who contributed to the art world's discourse on nonhuman agencies in the arts. This text is devoted to her with deep love and respect.

K-9_topology: HYBRID FAMILY by Maja Smrekar, cocurated by Jens Hauser and Jurij Krpan, Freies Museum Berlin, Bülowstrasse 90, 10783 Berlin. Studio visits January 29–February 2, 2016.

Raskin on "AI & Bio Art," Out of the Box—Ars Electronica festival, Linz, 2019. © Tom Mesič

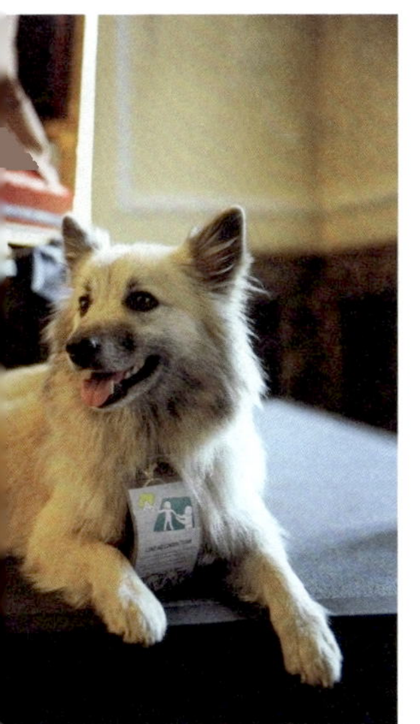

Digital Content 07 Maja Smrekar, *K-9_topology: HYBRID FAMILY*, 2016.

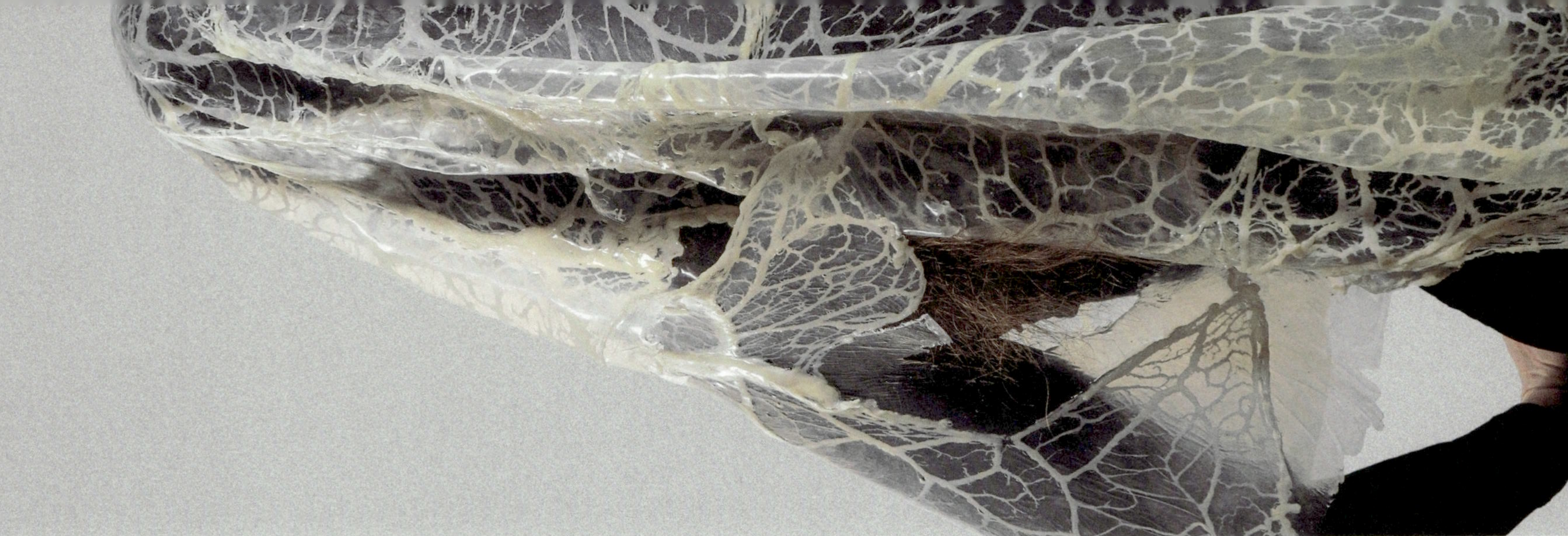

Re-Entangled Cohabitation: Immersed in Planetary Matter a New World is Unfolding. Can You Feel the Ecstatic Sensation of Alien Copresence?

Sylvia Eckermann and Gerald Nestler

Fig. 72 Sylvia Eckermann, *Untitled*, 2023, digital print. © Sylvia Eckermann

> What you see around you—in the physical space of the gallery as well as the virtual space from which I'm speaking to you—is our prototype of an exploration into speculative agency. To Yuk Hui's insight of a "cosmos that refers to locality," we'd like to add that cosmos speaks of the sheer abundance of entangled diversity. Can we "re-entangle" with it by grasping the virtual as a membrane, as a connective tissue for the myriad webs of life we inhabit—our planetary skins? Because cosmos is everywhere, and everywhere slightly differently.[1]

This statement made live at *Planetary Skins* briefly outlines the trajectory of the mixed reality art project. *The Future of Demonstration,*[2] which this project is part of, is an art series by the authors. It started as the City of Vienna's media art festival (2017 and 2018, in collaboration with Maximilian Thoman) and has explored vectors of mobilization since its beginnings. While *Planetary Skins* (Season 3, 2023)[3] dealt with the materiality of the real and the digital by focusing on the potentials, ambivalences, and conflicts induced by hybrid forms of presence, *Like a Ray in Search of its Mirror* (Season 4, 2024)[4] traced potentials for copresence through forms of appreciation that encourage more-than-human awareness and ecological solidarity. What ties these two projects together is their speculative

1. Excerpt from Gerald Nestler's "interludes" for the art project *Planetary Skins*. See "Planetary Skins," filmed at Galerie Thoman, Innsbruck, Austria, May 27, 2023, video, 1:23:27, accessed July 3, 2024, www.thefutureofdemonstration.net/planetaryskins/.

2. *The Future of Demonstration* is used as an artistic example in Gerald Nestler, "Contingent Claims: The Performativity of Finance, or How the Future Materializes in Technocapitalism," *Performance Research* 25, no. 3 (2020): 114–22, https://doi.org/10.1080/13528165.2020.1807770. The article and images of the performance *Making the Black Box Speak* were shown at the exhibition *Holobiont: Life is Other,* Magazin 4, Bregenz, 2021, and at the Angewandte Interdisciplinary Lab (AIL), Vienna, 2022.

3. *Planetary Skins,* a performance and exhibition shown at Galerie Klaus & Elisabeth Thoman, Heart of Noise Festival, Innsbruck, Austria, and online, May 27–June 3, 2023. See "Planetary Skins," *The Future of Demonstration,* accessed July 3, 2024, www.thefutureofdemonstration.net/planetaryskins/.

4. *Like a Ray in Search of its Mirror,* a performance and exhibition shown at the donaufestival, Forum Frohner, Krems, Austria, and online, April 19–28, 2024. See "Like a Ray in Search of its Mirror," filmed at Forum Frohner, Krems, Austria, April 19, 2024, video, 1:19:58, accessed July 3, 2024, www.thefutureofdemonstration.net/s4/stream.html.

exploration of planetary interconnectedness together with emerging cosmologies and cosmotechnics, and the application of post-disciplinary methodology, bringing together storytelling, conversation, performance, VR and other digital media, video, sound, installations, and other artistic, discursive, and technological tools.

One reason why we address ecologically, socially, and technologically connected themes is that alienation is now a planetary affliction with social and psychological consequences. Human beings act like aliens in environments that are vital to survival—and not just ours. Human impact is affecting more and more habitats, which are becoming hostile to liẏe. At the same time, our growing understanding of liẏe(-forms) is forcing us to recognize how closely linked natural processes and liẏing beings are. Evidently, the dichotomy between alienation and a/biotic entanglement cannot be overcome without radical change. How can we, as humans, come up and align with a new worldview? And how do we transform complex, often abstract insights into concrete, liẏed reality?

Change is urgently required, and we know it. But what kind of change? Will it be a humanistic reorientation, as we are often reassured? A new Renaissance? A new Enlightenment? Wouldn't they just keep us on the path of Western modernity, with its systemic colonializing grasp? Shouldn't we emancipate ourselves from such outdated concepts and the violence they continue to spread and instead listen closely to chthonic-cosmic *poets* and *makers* from both human and other-than-human worlds, whose bond to the Earth, the air, and all that liẏes can teach us about cohabitation?

As philosopher Fahim Amir succinctly put it, "Cohabitation does not prove that another world is possible, but that a thousand other worlds exist."[5] Thus, we should appreciate the wild and chaotic, the turbulent and contingent, the alterity within entangled planetary relations. And we should claim pluriform cosmologies that resolve humanism and its self-referentiality in the immanence of being-as-love. Philosopher Emanuele Coccia deplores that

5. Fahim Amir in "Like a Ray in Search of its Mirror," 00:39:00.

"[t]here are many new forms of experimentation of love in our society, and yet, there is not so much discourse about this."[6]

Artist and theoretician Patricia Reed opened the performative discussion of *Like a Ray in Search of its Mirror* by tracing how we make sense of the planetary:

> I think the general bond that we're going to be talking about a lot tonight concerns the important influence of narration upon human life worlds, in our triumphs, in our banalities, and, yes, in our atrocities. And to this end, the polymath Sylvia Winter has named us as "homo narrans," that is, hybrid, biophysical, and storytelling creatures who auto-institute their sociotechnical life worlds. And at this historical moment, we can now recognize ourselves in a planetary historical epoch.[7]

To Marian Kaiser (the MC), who asked what she meant by "planetary," Reed responded,

> I think what the planetary reveals to us is basically the real consequences of those artificial narrations upon earth systems, as these narrations have manifested in complex technological externalizations that intervene in the very Earth's system, sustaining the very possibility of human life. So, what the planetary sort of compels is a facing-up to an unprecedented historical material condition that requires us to practice coexistence in an environment in common, no matter how differentially it is inhabited. And I think the uncertainty of this moment sort of marks a juncture between two historical worlds or two historical epochs: to the unprecedented planetary forms of life that are now being invented, to planetary forms of life to come, and the end of a world as it's been

6. Emanuele Coccia, "On Plants and Planetary Names: A Conversation with Emanuele Coccia," interview by Joss Allen, *NO NIIN* 20 (October 2023), accessed July 3, 2024, https://no-niin.com/issue-20/on-plants-and-planetary-names-a-conversation-with-emanuele-coccia/index.html.

7. Patricia Reed in "Like a Ray in Search of its Mirror," 00:06:07.

known through the frameworks of Euromodernity and all its Janus-faced realities.[8]

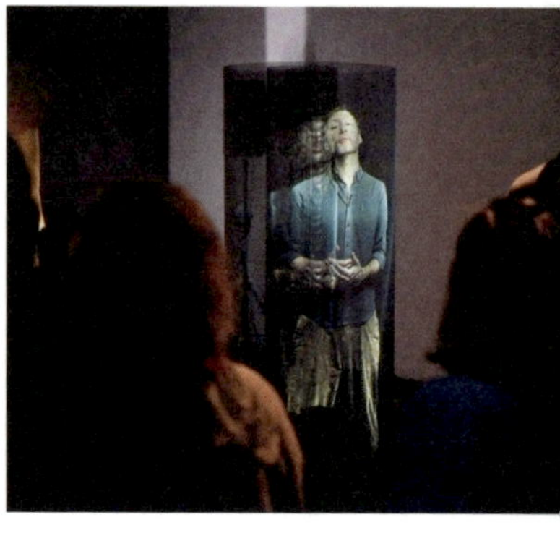
Fig. 73 Sylvia Eckermann and Gerald Nestler, *Like a Ray*

To address these themes in conversation, we developed an artistic, postdisciplinary format for *The Future of Demonstration* that attempts to meet the complexity and ambiguity, the atrocities and beauty at hand. *Postdisciplinarity* is a methodological "conception and framework in which art embodies risk and takes the leap from contemporary art's open promise to an aesthetic in the field of consequences that claims the present: a practice in which the *artist-as-collective* resolves aesthetics of critique into poetics of resistance."[9]

The figure of the *artist-as-collective* is an attempt to shift toward subjects composed of a "multitude of affiliation, alliance, assemblage, material as well as opponency and controversy."[10] When we accept a/biotic cohabitation as reality, "conceiving the individual as singular makes little sense, neither artistically, nor philosophically, nor politically. [It only] violently abstracts living assemblages and immixtures to capitalist segregation and extraction (including the art market and its capitalization of the artist brand)."[11] Rather, the idea is to "provoke works which are not objects of beauty for disinterested pleasure or interest-bearing investments, but collective subjects with their very own, *poietic*, agency in time."[12]

"Because action is only meaningful in context, *collective resolution* extends the conversation to wider collectives, audiences

8. Ibid., 00:07:01.

9. Gerald Nestler, "Renegade Activism and the Artist-As-Collective," in *Data Loam: Sometimes Hard, Usually Soft*, ed. Johnny Golding, Martin Reinhart, and Mattia Paganelli (Berlin: De Gruyter, 2020), 449, https://doi.org/10.1515/9783110697841.

10. Ibid., 436.

11. Ibid.

12. Ibid.

and allies."[13] Here, we make use of the semantic and semiotic field of *resolution*—ranging from perception, visualization, cognition, and knowledge production to problem-solving, decision-making, and joint determination (often including public action). Because of these interconnections, and because each term's meaning is accompanied by its own techniques, tools, and materials, we can conceive of resolution as a set of tools for generating transparency and collectivity. Resolution is not restricted to technical appropriation, such as visualization devices (for previously undiscovered worlds), big data analysis, algorithms that promote commodities, or the like. Nor is it merely a cultural technique of conciliation and consultation to craft compromise and compensation. Rather, its trajectory points toward openings and perceptions, and as such, resolution techniques are powerful tools for mapping and comprehending the liýe's dimensions, scales, interconnections, and hierarchies, its abundance and copresence. Resolution is therefore a methodological category, though not devoid of ideology, that can be utilized to assert claims to power. Technocapitalist resolution, including AI, has led to major advances but also massive asymmetries. Unfortunately, it has led to a development that "does not make us see and know; it makes us seen and known."[14]

In contrast, Gerald Nestler conceives of "resolution as a template to activate ecologies of solidarity,"[15] as a speculative method that can be applied to experiment with modes of cohabitation—and as a toolset to transform resistance from critique to new forms of insurrection. "What needs to be done is to radically exceed and reform the frameworks of critique and dissent, which, like transparency, evoke the false belief that we are all on the same plane."[16] In this respect, Daniela Gandorfer asked which roles law, governance, and blockchain technology should play and how they should be altered for that purpose:

13. Ibid.

14. Nestler, "Contingent Claims," 119.

15. Ibid.

16. Ibid.

> Thinking with and through breath, thinking with and through the atmosphere, through every single breath, human and more-than-human, ... each entangled with the others, shows us the limits and flaws of our Western legal ecosystems. So, we wonder, can we collectively and across differences work towards the decentralized concept of entangled and more-than-human rights? And if so, where does the power come from and how is it distributed?[17]

Philosopher Fahim Amir responded with an example that shifted the conversation to contaminated environments:

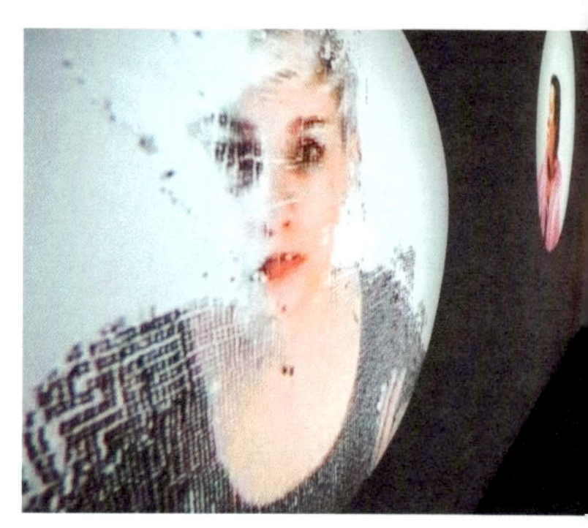

Fig. 74 Sylvia Eckermann and Gerald Nestler, *Like a Ray*

> When we're talking about atmospheres, I think it's important not to forget those who don't have a right to breathe. I'm speaking about, for example, Afghanistan. There's something called the Kabul cough. When you're three days in Kabul you start bleeding from the nose and must cough a lot. And that's because twenty years of a war of the West has left this country ruined, both the air and the ground. And I want to remind us all about the colonial legacies of the atmosphere. When a lot of countries were decolonized and freed themselves in the 1950s, one thing wasn't freed, and that was the atmosphere. The air above the ground was still part of the colonial framework for decades and decades after that.[18]

Amir's interjection is reminiscent of a story by Marian Kaiser, with which the media theorist and dramaturge illustrated the glaring

17. Daniela Gandorfer in "Like a Ray in Search of its Mirror," 00:12:19.

18. Amir in "Like a Ray in Search of its Mirror," 00:24:05.

violence and injustice of capitalist circulation and what matters to it as a/biotic matter:

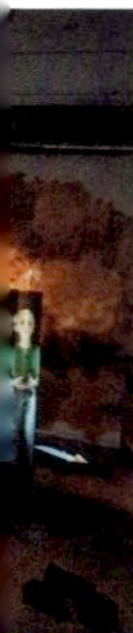

> A romantic red sky needs a sufficient amount of particles to refract the light. When a Chinese road work corporation recently started building overpasses to ease Kinshasa traffic, they had to strip up to nine meters of soil before they hit stable grounds uninfused with plastic fossils—those archaeological remnants of Northern industrial capitalism now piling up in simmering roadside heaps. ... Extractive forces digging up prehistoric beings and mythical creatures that have slowly rotted into fossil fuels over millennia, burning them or hardening them into colorful plastic forms to be sent out over the planet to be burnt a little later somewhere else, dissolving into colorful skies or to return into the ground in places they were first dug up from to be unearthed yet again by ... underpaid Congolese workers under most horrible conditions they for sure did not create.[19]

Kaiser added, "unfortunately, the virtuality of this reality has not taken away any of its realness. That is, if we don't mistake virtuality for something happening in some ominous digital sphere behind the screen of some of electronic gadget separated from our 'real' world."[20]

Exploitation is pervasive physically and digitally, and our entire liyeworld is becoming fragile in its wake.[21] Do we lack the imaginative power to make the leap from our segregating, marginalizing, self-limiting humanism to wild and openly aware sensing and sense-making? To imagine risk, for instance, not as a burden or trap for others, but as a space of possibility and self-determination? What about the courage to imagine something that "decenters"

19. Marian Kaiser in an unpublished text written for "Planetary Skins." Kaiser liyed in Kinshasa, Democratic Republic of Congo, at the time.

20. Ibid.

21. The term "wake" needs to be understood in all the registers that Christina Sharpe conveys in her radical, bold, poetic refutation of white supremacy. Christina Sharpe, *In the Wake: On Blackness and Being* (Durham: Duke University Press, 2016).

us, leading to the joyful appreciation of what exists and liýes, and allowing us to turn from seeking the next fixed position—the gridlock we call "security"—to a movement more akin to breathing, a sense of being in and with the world, a resolute shift toward inclusive activation that is not afraid of what circles back? The hard question, of course, is how to turn desirable ideas into a liýed reality that people accept. As Gandorfer said regarding her field of research:

> How do we reconcile legal frameworks rooted in sovereign states with the emergent realities of decentralization? Can we conceive of a decentralized right to breathe, one that addresses the nuanced injustices inherent in every breath while acknowledging its intrinsic value? ... Merely critiquing the status quo falls short of our imperative to creatively engage with the emerging possibilities, mobilizing our complicity towards novel configurations and an appreciation of diversity and difference.[22]

We are agents in nature's composition. We are learning this lesson the hard way, and opposition to it is still disproportionate. Instead of turning a blind eye to air pollution, global warming, mass extinction, dying oceans, and other consequences of extractivism and exploitation, we should seize the opportunity to learn how other cosmologies can teach us to become sentient co-actors—and, in symbiosis with others, to grasp what it means to coinhabit an atmosphere of interspecies copresence.

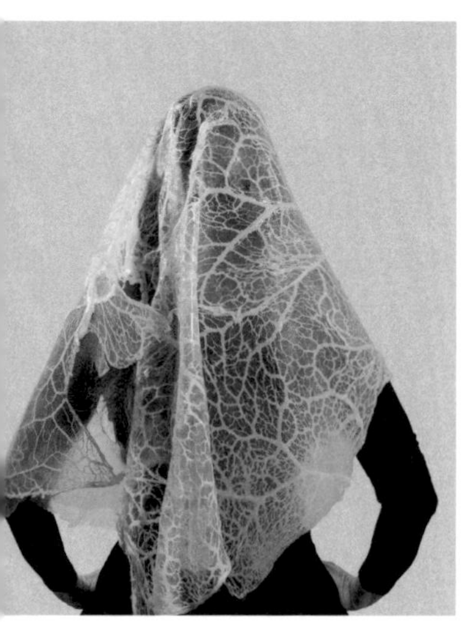

Fig. 75 Sylvia Eckermann, *Untitled*, 2023, digital print. © Sylvia Eckermann

22. Gandorfer in "Like a Ray in Search of its Mirror," 00:10:41.

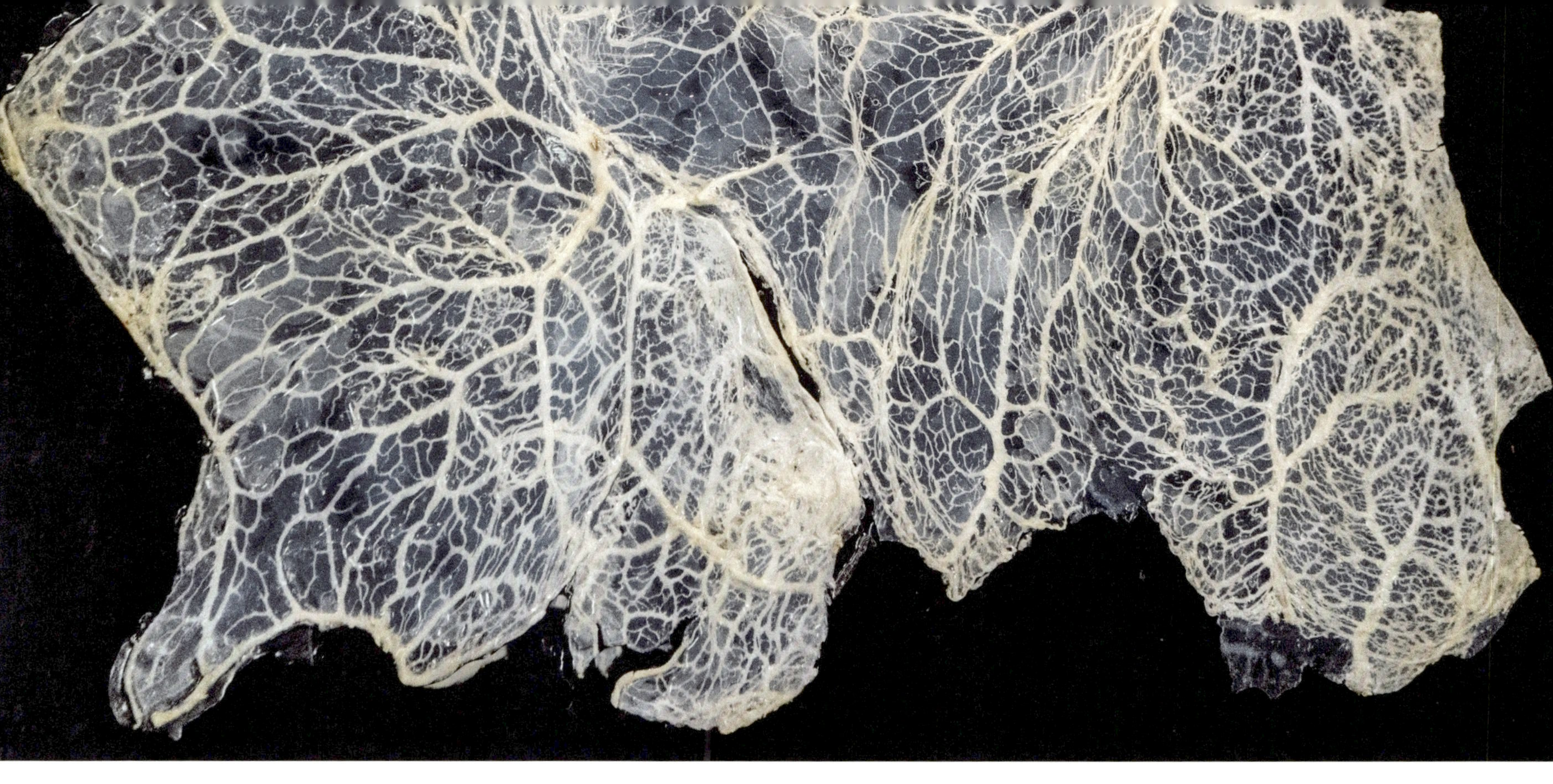

Fig. 76 Sylvia Eckermann, *Untitled*, 2023, digital print. © Sylvia Eckermann

“To understand a climate is to grasp an atmosphere,” says Coccia.[23] To “grasp” is to take something “other” in your hand; symbolically, it transforms the hand into thought, touch into vision, sensing into sense-making. In an intimate exchange of affect and effect with an “other,” grasping activates deep connection and manifests the self. If it directly affects the connection between body and mind, isn’t our grasp our first tool?

We exist with and because of a/biotic others. The ongoing mass extinction of species is happening not only outside but also inside our bodies. Our “grasp reflex” is decimating liƴe-forms that have kept us aliƴe and breathing symbiotically for millennia. Each body is a cosmos in symbiosis with myriad others, all concurrently local and universal in “our interconnectedness with the biosphere.”[24] As space architect Barbara Imhof related in *Planetary Skins*:

> There is the insane number of about 30 trillion microbes in us and around us. Just on the skin there are about as many as there are human cells in our bodies. ... If we were microbes and would live in this micro-dimension, populating and experiencing ourselves in our body, our body would have about the length from here to the moon and back. ... [T]he human body would be a gigantic planet on which we would live.[25]

Our bodies are volumes of differing magnitudes made up of innumerable entangled scales within a collective liƴing being. It depends on the resolution inhabited whether the human body is perceived as a soft and solid whole, as humans do, or, as a vast region to traverse and nest in, as microbes might do. Artist Thomas Feuerstein extrapolates this thought from the human body to the Earth: “Around 70 per cent of all bacteria and archaea do not live in the sky [or] in sun-flooded paradise gardens, but in Hell, in and under the epidermis of the Earth. ... The diversity of life does not

23. Emanuele Coccia, *The Life of Plants: A Metaphysics of Mixture* (Cambridge: Polity Press, 2019), 118.

24. Quoted from the blurb of this book.

25. Barbara Imhof in “Planetary Skins,” from 00:41:15.

originate on the surface; rather, it grows in the deep pores of the global skin."[26]

How radically apocalyptic, rather than abject and liminal, will liyfe on Earth turn out to be? Will our others continue to exist together with humans, or will nature shape liyfe on Earth without us? This question, of course, affects animal liyfe, too. As Amir observed:

> We profoundly lack the concepts to be able to grasp and articulate the extent and nature of the presence of animals in the world. To include the claims of others as part of the equation is to relativize our sovereignty. Participation also means sharing, and so we will have to give up something. For instance, the idea of a completely controllable and "manageable" environment. All life forms inhabiting the planet influence each other in complex ways that are neither fully understood nor controllable. This also means welcoming "unintended landscapes" as encounters with nature, landscapes that have not been created or designed with a purpose. Such enclaves of disorder are habitats for unexpected forms of sociality.[27]

And, Amir argued, "We need to find more imaginative languages, more apt images, and more inspiring models that help us to better perceive and understand this new world in which we have long been living."[28] We need to cultivate a worldview where humans recognize themselves as part of nature. This change needs to be based on a transition from the one-dimensionality of the alienating global paradigm to a shared diversity of planetary liyfe.[29] It is linked to the realization that liyfe permeates all dimensions of nature. Science and visualization technologies are making the immeasurable

26. Thomas Feuerstein in "Planetary Skins," from 00:14:34.

27. Fahim in "Like a Ray in Search of its Mirror," 00:36:36.

28. Ibid.

29. Gayatri C. Spivak introduced the term "planetary" in 1997 to counter the globalized alienation produced by capitalism. Gayatri C. Spivak, "Imperative to Re-Imagine the Planet," in *An Aesthetic Education in the Era of Globalization* (Cambridge, MA: Harvard University Press, 2013), 335.

"interdependencies of animate and inanimate realms"[30] in which humans are involved increasingly tangible.

For us, Fahim's appeal also calls for new forms of collaboration between art, science, and technology to counter technocapitalist fantasies of AI as the singularity as well as the metaverse and other predictive simulations that are once again producing asymmetries and biases, widening the gap between rich and poor, exacerbating carbon footprints, exploiting divisive politics, and reestablishing reactionary norms of class, race and gender. There is no way out—we are part of nature's relational processes. But there are ways in, with many forms of sensing and sense-making to explore. Postdisciplinary approaches can anchor this wealth in our imaginations and make it fruitful in practice.

The techno-ecological format of the artistic interplay between *Planetary Skins* and *Like a Ray in Search of its Mirror* reorients the concept and aesthetics of VR. While conventional VR and metaverse projects are invested in creating artificial 3D worlds and simulating reality, we transform the virtual into an "osmotic skin" that connects physical and virtual space. Our earlier project, *Planetary Skins,* prototypically experimented with point-cloud video to first integrate participants into our VR environment (in the Unity3D game engine) and then into the physical gallery space (via projections).[31] In *Like a Ray in Search of its Mirror,* we used our novel technology to replace VR avatar representation entirely. Real-time point-cloud transmission liýe-streamed the appearance, movement, voice, and body language of human and nonhuman participants, producing an intensive space of emotional

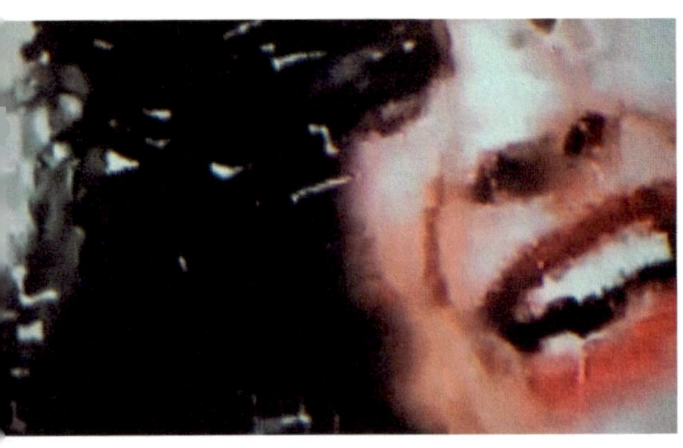

Fig. 77 Sylvia Eckermann and Gerald Nestler, *Like a Ray in Search of its Mirror,* 2024. © Michael Loizenbauer

30. From the abstract for *Liýe is Other.*

31. Please see the project's credits for the team of artists, architects, and technicians that made this artistic experiment possible.

connection.[32] Now, the virtual is a membrane through which people and other beings can directly interact, talk, and perform with each other from wherever they are.

By connecting physical and virtual worlds through art and technology, we aim to deepen mutual perception and collaboration by means of a wide range of experiments that include people, other beings, things, and artworks. Contributors are present in other ways, too—for instance, projected onto metal spirals, art objects that create a holographic effect.

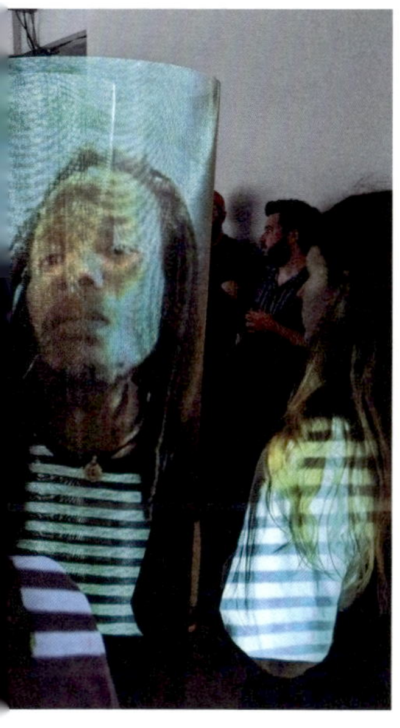
Planetary Skins, 2023. © Thomas Thaler

These novel technologies and aesthetics are embedded in the artistic environments of *Planetary Skins* and *Like a Ray in Search of its Mirror*. They incorporate other materials and media, and thus lend a voice to other forms of artistic expression. These spaces were created by Sylvia Eckermann, whose installations use an/organic materials that define the gallery space using patterns found in nature in a wide variety of forms and applications, from microorganisms and mycelia to insect wings and leaf veins, from network structures and synapses to cosmic supernovae. Martha Laschkolnig and Markus Passecker's

32. The voices include fungi, microbes, plants, animals, inorganic and synthetic matter, technologies, tools, and humans: Giuliana Furci (live from Santiago de Chile) delved into the networks of mycelium ecosystems. Fahim Amir discussed cohabitation by looking at the unruly politics of human-animal relations. The rewilding of Waldrapp birds was outlined by Gordan Savičić and Felix Stalder. Maggie Roberts (0rphan Drift, live from London) contemplated the nine brains of the octopus and its wildly alien sensing. Martina De Dominicis and Alberto Cissello (debocs) explored sensations of touch. Agostino Nickl prompted abstract neural network constellations that mirror our collective imagery. Thomas Nail (live from Denver) reconsidered the motion and swerve of matter, and its porous constitution. Yuka Takahashi (live from Columbus) performed her point-cloud body. Daniela Gandorfer (live from London) speculated on more-than-human governance. A different sense of planetary space and inhabitability was imagined by Patricia Reed. Based on lyrics by Gerald Nestler and Volkmar Klien, Wientaler Dreigesang and Soulcat E-Phife composed sound for philosophical and political resistance. Marian Kaiser took up threads to weave a tapestry from all our artistic-discursive encounters. And the audience took part either on site or online.

biofeedback performance on Lucretius's poetic treatise on matter, as well as Agostino Nickl's constellations compiled using CLIP, an OpenAI pretraining tool for artificial intelligence, were juxtaposed with a video of scientific micro and macro images by Eckermann and Nestler. Together, they manifested the significance and complexity of the concept of *cosmos*.

Such aesthetic qualities allow us to engage with various forms of performative participation and representation. Interactive and mediated copresence reinforces the immersive liyfe character of mixed-reality art projects for participants and the audience. And while its intensity and volatility make the consequences of alienation palpable, it also creates a common space for a/biotic encounters with a diversity of practices, ancient and indigenous, animated and artificial, scientific and cultural—methods of learning and unlearning that can reconcile us with the fact that we are part of nature's wild entanglements. In fact, this involves derivative speculation[33]—engaging in experimental flights and recalibrations of self-transgression, instead of being fundamentalist at the core. Derivative speculation is less about fixed objects and objectives because it focuses on, and immerses itself in, re-entangling with ephemeral and volatile relations—embodied matter whose emergence incessantly weaves reality.

Alienation, as we know, not only refers to capitalist production and class relations but also bears on humans' relationship with nature, something that Karl Marx grasped, referring to the "metabolism" between our inner nature and the nature that surrounds us. Marx recognized the rift[34] in that metabolism caused by capitalist dynamics. The consequences of a growth-based, global political economy are being felt throughout the world, from soil depletion (which Marx mentioned) and the exploitation of resources and labor to species extinctions and climate catastrophe. More alienation is hardly conceivable.

33. Nestler, "Contingent Claims," 114–20.

34. Drawing on Marx's earlier work, sociologist John Bellamy Foster introduced the term "metabolic rift" in his 1999 article, "Marx's Theory of Metabolic Rift: Classical Foundations for Environmental Sociology," *American Journal of Sociology* 105, no. 2 (1999): 366–405.

Aiming to explore transformative potentials, we experiment with re-entangling as a speculative shift—not that humans are completely disconnected from nature. More and more people are recognizing our destructive influence and the need to change course. Timothy Morton speaks of "hyperobjects"—"there is no center and we don't inhabit it. Yet added to this is another twist: there is no edge! We can't jump out of the universe. ... We're always inside an object."[35] Yuk Hui focuses on cosmotechnics: "'Cosmos' does not refer to outer space, but, on the contrary, to locality."[36] The term *cosmos* designates a diversity of geographies as well as imaginative wealth and material abundance. Making recourse to Arthur C. Clark,[37] we connect to its alien magic by working with the virtual—against the prevailing use of the term—as a magical technology that offers resolution as a planetary membrane mediating between (often conflictual) geographies and imaginations but also a cosmic tissue connecting countless networks of liƴe and matter—dimensions we (co-)inhabit, despite our self-centeredness.

It was at this (conceptual) point that Patricia Reed imposed a caveat. And because other participants joined the discussion and shifted the perspective,[38] the following is a longer extract from the performance, mapping the course of the conversation:

> Reed: I'm a bit worried about importing tropes or discoveries in the physical world [into] systems of governance. ... Because it's also been this kind of problem of a natural fallacy of importing terms like entanglement and assuming they have some kind of moral predisposition upon our artificially constructed worlds.

35. Timothy Morton, *Hyperobjects: Philosophy and Ecology after the End of the World* (Minneapolis: University of Minnesota Press, 2013), 17.

36. Yuk Hui, *Art and Cosmotechnics* (New York: e-flux, 2021), 40.

37. The third and final of "Clarke's three laws" reads: "Any sufficiently advanced technology is indistinguishable from magic." See "Clarke's Three Laws," *Wikipedia*, accessed July 3, 2024, https://en.wikipedia.org/wiki/Clarke%27s_three_laws.

38. Alongside Kaiser and Amir, philosopher Thomas Nail, mycologist Giuliana Furci, artist Maggie Roberts, and media theorist Felix Stalder.

Amir: I'm not an expert in quantum mechanics and questions of entanglement. But I would in defense of this position argue that at least [on] the left it was traditionally ... that in manifestos you detect parts of reality that you consider good and progressive, and you concentrate and focus on that. And I think this is a materialist perspective on the world, not just great ideas.

Thomas Nail: I agree, and one of the tricky things is that there's no shortcut. The politics and all our ontologies, ... [be] they based in quantum physics or elsewhere, they're all historical. ... [T]hey are practices, performances that we engage in, that we make. And there's no shortcut to the experimental process. Just because everything's in relation doesn't mean everything's going to turn out okay.

Giuliana Furci: Fungi are organisms that live inside their food, contrary to us, who have food outside of our bodies. So, the duality between the ways we live and the way that we understand our environment are completely different. There is no difference between the outside and the inside in fungal life because they graphically demonstrate that no one is without another, individuals don't exist.

Maggie Roberts: There's no inside or outside for octopuses [either]. ... I saw this thing once, and it was as if an octopus was everywhere in the world, and we wouldn't know that because it camouflages itself [to] everything like a kind of protein skin. So, it becomes the surfaces of everything, ... and I can't tell if everything's in fact octopuses.

Kaiser: Thomas, would you say that the conditions we live in or under or with, that we are a lot closer to myceli[a] or fungi or octopus[es] than we might think?

Nail: Yes, I would say that. But it's tricky because we live in a state ... of constant performative contradiction. [For]

> [m]illions of years we've evolved this hypersensitive organism you know our bodies, our nervous system. We really are, or can be, deeply connected to all the things around us and very much continuous with those fungal pores and octopus feeling[s]. But we live in a world which constantly tries to cover that up with walls and cement and colors and uniform patterns. ... We've lost huge amounts of these iterative fractal patterns that used to shape our consciousness. And now we do our best in industrial civilization to pretend that the natural world doesn't exist. That we don't have to be attuned to it in the same way.
>
> Felix Stalder: Yes, but this also opens a paradox. These worlds exist, but they also need to be brought into being by opening towards other species, not just for connection and entanglement but for a very precise goal: for increasing others' autonomy. And this poses a difficult challenge for Western thinking that for science and technology has always been about prediction and about control. But we have to realize that we are not in control of this shared destiny and that our agency should not increase necessarily the control over others. So, what we need is a new type of mastery—the mastery of non-mastery.[39]

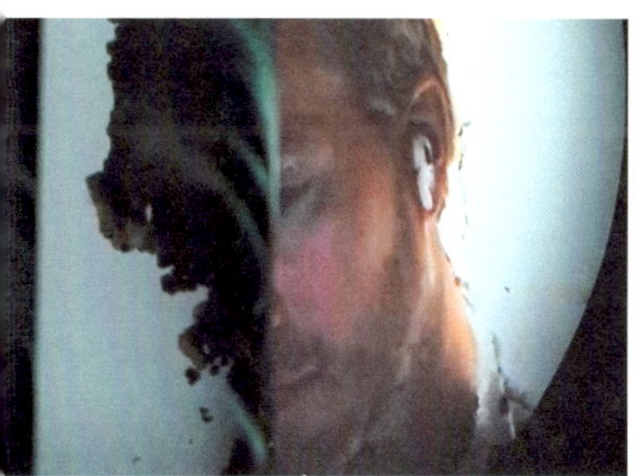

in Search of its Mirror, 2024. © Michael Loizenbauer

In *Planetary Skins*, quantum physicist Tania Traxler elaborated on the science that gave birth to the terms *entanglement* and *indeterminacy*. And she saw potential for a materialist perspective for creative, yet sound, transdisciplinary interpretations:

> [The] physical vacuum hints at the fact that the boundary of possibility and reality is not as sharp as long presumed. The common view regards possibility and reality as [a] dichotomy,

39. Reed, Nail, Furci, Roberts, and Stalder in conversation in "Like a Ray in Search of its Mirror," from 00:15:29.

> according to which something is either real or possible but never both at once. … However, if [the] vacuum holds the inevitable possibility of harboring virtual particles, this indicates a mode of possibility which is real in being possible. Virtual particles are real in a fundamental sense, they are not limited to mere mathematical possibilities, but hold real consequences without being materialized. … [Q]uantum field theory tells us that virtual particles act as … messenger[s] between other particles. Virtual particles embody what philosophers like Henri Bergson and Gilles Deleuze dreamed of: a form of possibility which is real in its being possible.[40]

For Nail, there was an ancient antecedent to quantum theory, namely Roman poet Lucretius's conception of matter in his treatise *De Rerum Natura*: "For Lucretius, nothing is static. Everything is migrant and mutational. The movements of matter cannot be fully determined, it has a kind of irreducible agency, which continually weaves and unweaves the visible universe around us. … Scientists call this 'metastable states.' At every scale of reality, energy flows, and matter cycles, and the energy is released. … There are only processes and patterns-in-motion."[41]

In order to form hybrid cooperations between "planetary beings,"[42] our contributors are humans as well as fungi, plants, animals, compounds, materials, and technologies. Our art projects concern liying matter, other alien communities, and transitory zones of liyeness. These ideas relate to a wild, "mad" imagination, which we see in philosopher and artist Johnny Golding's philo-poem, "Octopussy thinks on the problem called matter":

> Not for a minute did he worry about being outside (or inside) the world. They were world; they were matter and that was all that mattered. She knew of wetness. She knew of appetite. She even knew of ambush, camouflage, battle, and now, even

40. Tania Traxler in "Planetary Skins," from 00:52:09.

41. Thomas Nail in "Planetary Skins," 00:09:08–25 and from 00:48:46.

42. Gayatri C. Spivak, "Imperative to Re-Imagine the Planet," 335.

> boredom, as somehow entailing matter whilst being matter. She knew of sleepiness, pleasure, pain, even jealously. She was quite well aware of scale, dimension, and, indeed, size.[43]

In Golding's poem, diversity forms a multifarious being-in/as-ecology. In a recent article, Coccia puts forward the provocative thesis that "to think about the planet, [it is] better to make art, not ecology."[44] And he reasons in a mode akin to what Nestler describe above as subjects with their very own, *poietic* agency in time (the *artist-as-collective*):

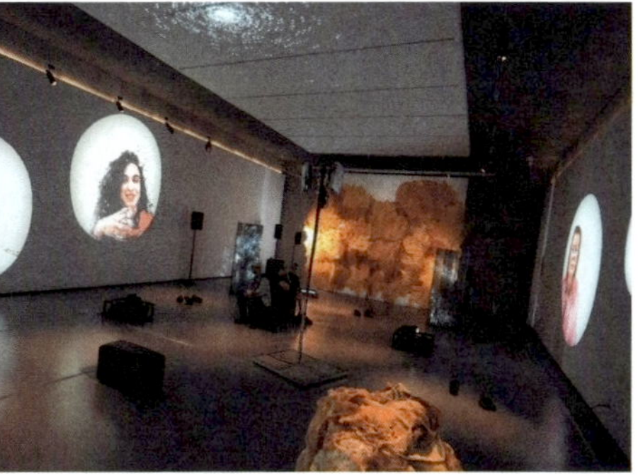

in Search of its Mirror, 2024. © Michael Loizenbauer

> Knowing Earth as a subject means knowing it through art. But it is not that art has to represent Earth, nor is it enough to bring plants into a museum to know the planet. It is about relating to Earth as we relate to art; that is, thinking of the planet itself as an art form, as a kind of open-air museum for contemporary nature. ... Earth must become a planetary theater: not in the sense of a building, but in the sense of a space in which everything—plants, animals, lichens, fungi, stones, winds, clouds, etc.—is perceived as an actor, as capable of acting, as a subject.[45]

Nothing is unnatural, not even artificial intelligence. Nature is everywhere, and all matter exists in relation to others. "Natural" is a construct. Technology is part of our cosmic ecology, from the depths of rare earth to the shiny surfaces of user interfaces. *The Future of Demonstration* offers a discursive-performative format in

43. Johnny Golding in "Planetary Skins," 01:12:52.

44. Emanuele Coccia, "Don't Call Me Gaia," *e-flux Architecture*, October 2023, accessed July 3, 2024, www.e-flux.com/architecture/hydroreflexivity/571453/don-t-call-me-gaia.

45. Ibid.

which sharing ideas, experiences, and knowledge is so concrete that constellations of planetary copresence become tangible.

How do we deepen the resolution of our knowledge and what we feel and perceive? Can techniques of resolution—optics of sensing and sense-making—act as a collective toolset, as a methodology to chart and forge yet unknown alliances with beings that are often more sentient than Euromodern thought has led us to believe? Is this the postdisciplinary "grasp" we need to acquire? And is this the insurrection we should be aiming for: disentangling ourselves from the "frameworks of externalization" and instead re-entangling with others by creating subtle new intimacies?

Coccia sees art as a modern form of animism because, in our encounters with it, "we accept the idea that it contains a psychological, emotional, or mental intensity that transcends the mere material reality in which it exists."[46] For us as artists, however, art now needs to dig deep into a/biotic material reality. We thus view art as a postdisciplinary practice that can interact with scientific disciplines and other imaginative knowledges, and can experience various worlds precisely because it is not subject to any disciplinary boundaries. Art becomes cosmic, speculatively probing into deep planetary localities that interconnect with others in nature's vast expanse. If Coccia is right, and art does remain animist, it will be despite its chthonic-chaotic embrace of contingency, which allows it to feel (with) others, between bodies, to explore matter that matters, and hence to *make* sense.

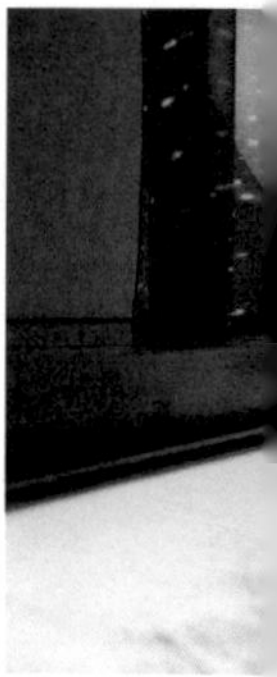

What becomes *aliÿe* here is the *artist-as-collective*: diverse species, compositions, and compounds coming together from different strata, continents, and ecosystems to activate sensing and sense-making as modes of resolution. And as a more-than-human association, we speculatively explore the resourceful potency of our myriad ecological, cultural, and technological relationships; or, if you like, the myriad planetary skins we inhabit in varied but interrelated ways.

46. Ibid.

We would like to end by returning to the first of our two projects, *Planetary Skins*. In her introduction to the live performance, Sylvia Eckermann described our approach as follows:

> *Planetary Skins* is like an "osmotic skin" in which the virtual mutates into a membrane to allow simultaneous presence in physical and virtual spaces. We use technology to exchange ideas across space and across time zones and to think about synergies together. In the words of the evolutionary biologist Lynn Margulis, we think of the world as an organism—as a system made up of organisms and matter that cannot be completely separated from each other. They are in close, often symbiotic relationships with one another through myriad interconnected and communicating layers. And we human beings are part of it! This approach can lead us towards a new kind of agency. An agency that learns from and joins forces with nature. An agency that moves from an obsolete humanism to celebrating the diversity and richness of life and nature![47]

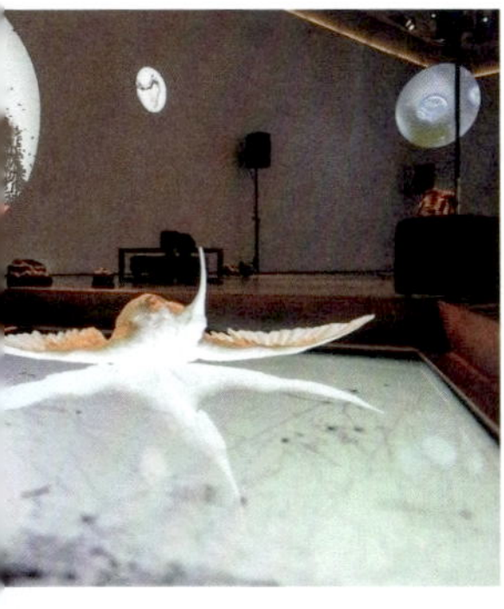

in Search of its Mirror, 2024. © Michael Loizenbauer

Can you already feel the ecstatic sensation of alien copresence? ∞

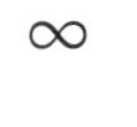

Digital Content 09 Sylvia Eckermann and Gerald Nestler, *The Future of Demonstration.*

47. Sylvia Eckermann in "Planetary Skins," from 00:00:20.

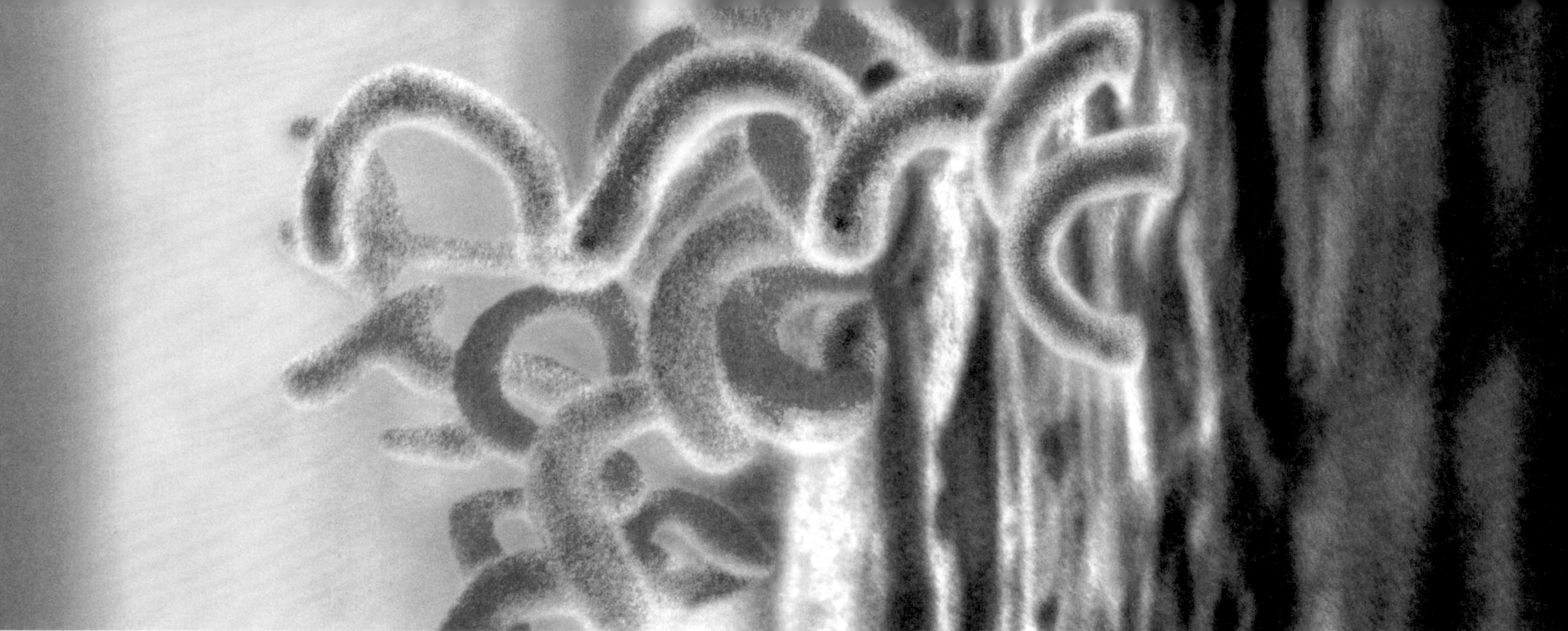

Art and Liyfe, or Carving the Atmosphere

Dorion Sagan

Fig. 82 Scanning Electron Micrograph (SEM) of *Treponema pallidum*. © Phanie Sipa Press, Alamy

Part One:
A Crash Course in Liyfe as a Work of Art

> It is something to be able to paint a particular picture, or to carve a statue, and so to make a few objects beautiful; but it is far more glorious to carve and paint the very atmosphere and medium through which we look, which morally we can do. To affect the quality of the day, that is the highest of the arts.[1]

This quote by Henry David Thoreau—a proto-ecologist, and the first to observe succession in ecosystems—combined with a similar famous sentiment expressed by Nietzsche—that "Existence is only ever justified as an *aesthetic* phenomenon"[2]—presents us with a pretty high bar.

I would like to "carve and paint the very atmosphere and medium through which we look," as Thoreau says. In fact, you may be surprised to learn that some entities that you likely consider your deeply invidious inferiors have already done so. I speak of scum. Specifically, of germs and bacteria. Independent journalist Valerie Brown, who began vocational liyfe as an artist and musician, made this brilliant synoptic comment: "[W]e eukaryotes are bacteria's art form—maybe not their *magnum opus* (that would be the entire planet) but at least their master's thesis."[3]

To see art in this broader context, we must greatly expand our notion of it and grant (let us not be too stingy) little purposes, compounded over billions of years of cosmic time on Earth alone,

1. Henry David Thoreau, *Walden* (New York: The Macmillan Company, [1854] 1910), 117.

2. Friedrich Nietzsche, *The Birth of Tragedy Out of the Spirit of Music (Die Geburt der Tragödie aus dem Geiste der Musik)*. The comment appears in the section "Attempt at Self-Criticism ... written for a new edition of *The Birth of Tragedy* in 1886, as the 'provocative' or even 'offensive' proposition that 'only as an aesthetic phenomenon is the existence of the world *justified'* (with Nietzsche's own emphasis [...]). That wording does not, in fact, correspond exactly to either of the two main formulations of the point in the original book, the first of which, in §5, states that 'only as an *aesthetic phenomenon* is existence and the world eternally *justified,'* and the second, in §24, that 'only as an aesthetic phenomenon does existence and the world seem justified.'" Stephen Halliwell, "Justifying the World as an Aesthetic Phenomenon," *The Cambridge Classical Journal* 64 (2018): 91–112.

3. Valerie Brown, January 12, 2012, private communication upon death of Lynn Margulis (1938–2011).

to the sensing, symbiotic, and group-living beings we have long derided as microbes—including the cells that evolutionarily constitute us. Their work, as minuscule as it might seem on an individual basis, is not of course just decorative, but a matter of survival. With their media the elements of the periodic table, they (who are themselves constituted by those very media) have assembled themselves from cosmically common carbon, oxygen, nitrogen, phosphorus, sulfur, and hydrogen atoms, the main stuff of stars, which themselves come in multitudinous and multifarious combinations. We are accustomed to thinking of things from a human standpoint, measuring them in Protagoran terms, so to speak, but from a broad, cosmic, evolutionary perspective, we are the epiphenomena of multi-billion-year-old and perhaps cosmically distributed, highly durable, and intensely social microorganisms. To take one example, stromatolites—domed rock formations built by sun-loving cyanobacteria slipping out of their polysaccharide sheaths to get closer to their local stellar source of photosynthesis, our local yellow star—are found not only in liẏing forms, for example, in Shark Bay, Australia, but in fossil forms going back to the Archaean Eon billions of years ago. These communities, which include other forms of bacteria living off the local largesse, have been called "bacterial skyscrapers"[4] and anticipate modern cities; similarly, we ourselves have been described as walking buildings made by bacteria.[5] Such descriptions, only partly hyperbolic, emphasize the functional necessity of liẏe as a mutual art form feeding upon and protecting itself from a highly energetic, complex, material manifold for trillions of generations, only a minute fraction of which have been human. Thus, the broadest view of art includes all liẏe

4. See, e.g., Dorion Sagan, *Livro de Seres Invisiveis* (Rio de Janeiro: Dantes, 2021); Lynn Margulis and Dorion Sagan, *Microcosmos: 4 milliards d'années de symbiose terrestre* (Paris: Éditions Wildproject, 2022); Lynn Margulis and Dorion Sagan, *Microcosmos: Four Billion Years of Microbial Evolution* (Berkeley: University of California Press, 1997).

5. As Nobel Laureate Richard Roberts, Chief Scientific Officer of New England BioLabs, says in the documentary, *Symbiotic Earth: How Lynn Margulis Rocked the Boat and Started a Scientific Revolution*, "You know, there are many people who think that they actually invented us so they had a nice place to liẏe. They're just like us building houses. The bacteria built houses. We ended up being them." *Symbiotic Earth: How Lynn Margulis Rocked the Boat and Started a Scientific Revolution*, directed by John Feldman (2018; Oley, PA: Bullfrog Films, www.hummingbirdfilms.com/symbioticearth).

as well as the human artist with or without delusions of personal grandeur. In contrasting this description of a general art with human art, we might mark two major differences: First, the art of liýe is never static, but exemplifies rather than merely demonstrates flow (as in, say, the paintings of Kandinsky)—a continuous, three-dimensional flow whose open frames are bodies themselves. Second, the necessity of using energy, and producing waste, in the art of liýing systems leads to works that incorporate that waste via distinct forms of metabolism. Thus, as processes, the artworks of liýe, unlike those of artists, are never finished. In addition, as part of the project of de-anthropocentrization, we should be generous as to the phenomenological "insides" of what Lynn Margulis called our planetmates—the others whose space we share and whose shared space, indeed, we "are."[6]

As James MacAllister asked in 2014,

> [W]hat if our consciousness—my I-ness—is just the job that the consciousness of my 90% bacterial cells and their obligate 10% animal multicellular cells require "me" to do? I have no consciousness of what all my cells are doing. I am like the super of an apartment building, I hear the complaints (e.g., an ache), I have to feed the furnace, put out the garbage, sweep the hallways, but I have no experience of the lives of the apartment dwellers or what they are really up to. My cells have lives and sentience of their own just like the apartment dwellers. Isn't this disconnection of awareness between scales of "ecosystems" fascinating? Maybe not disconnection but limited connection—sort of like our normal perception of reality that simplifies or filters out the dazzling truth of reality that is too complicated, messy or distracting for us to deal with all the time. It makes my consciousness sort of like our

6. The study of the experience of the others—prokaryotes and eukaryotes—of which we are evolutionarily composed, might be understood under the rubric of "biophenomenology." See Dorion Sagan, "From Empedocles to Symbiogenetics: Lynn Margulis's Revolutionary Influence on Evolutionary Biology," *Biosystems* 204 (June 2021), https://doi.org/10.1016/j.biosystems.2021.104386.

> corporate media whose job it is to misinform, distract, dumb down, manipulate.[7]

Eukaryotes are cells with nuclei in them. Almost always, they contain mitochondria, oxygen-using organelles with their own DNA, outside the nucleus. Often, they too have colorful plastids—green in plants, purple and brown in kelp and seaweeds—which also contain their own DNA. The organelles, coloring the cytoplasm outside the nucleus with green, red, and purple, were once independent liyfe-forms. Inside the nucleus, chromatin-bound DNA clustered into chromosomes comes from an ancient host, probably an archaeon, a type of prokaryote. The mitochondria and the plastids also come from prokaryotes, but their ancestors were true bacteria with stretches of RNA distinct from the presumably older prokaryotes, the archaea. You are not just an artist (if you are) but a eukaryotic artistic creation, because your cells (most of them) have nuclei. But there are plenty of other art forms to choose from: a single-celled *Amoeba proteus*, an *Amanita muscaria* mushroom, a wrinkled green pea, or a strutting, squawking peacock, fanning a nacreous sea of misplaced faux, azure-irised eyes.

All plants, fungi, and animals are eukaryotes. So are members of the motley kingdom sometimes known as protoctists, consisting of thousands of species, mostly unknown because they neither harm us nor offer any obvious potential medical benefit. There are, for example, 60,000 extinct species of foraminifera. And foraminifera have artistic tendencies.

Foraminifera, or forams as they are affectionately known, compose themselves in part from silica, silicon dioxide, sand. In experiments, they have been shown to choose one size and color glass bead over another to make their liyfing shells. The bodies of such beings often harbor other beings, photosynthetic radiolaria that can be as beautiful as liyfing cathedrals, reflecting and refracting the sun's light. The way forams they choose one substance over another to make themselves is reminiscent of the meticulous

7. James MacAllister, personal communication spurred after reading Dorion Sagan, *Cosmic Apprentice: Dispatches from the Edges of Science* (Minneapolis: University of Minnesota Press, 2013).

craftsmanship of artisans and of our own seemingly self-generated choosing behaviors, our undeconstructedly "purposeful" deliberations. Unlike most protoctists, foram species are well-studied because specific species demarcate oil deposits. But they are far more than functional in this narrow, human, industrial sense.

Martin Brasier (1947–2014), an Oxford paleontologist with a wonderful eye who found fossils in the stones of the dining hall of Balliol College and was once warned on a ship to stop doing so much work as he was making the rest of the crew look bad, presented a giant projection of the Sphinx at Giza during a memorial to my mother, biologist Lynn Margulis (1938–2011). Brasier pointed out that the Sphinx, that half-man-half-cat, imaginary being, was composed of real chimeras. They appear as nummulites, "coin-stones," in the anthropo-feline's rock-hard body. Greek historian Herodotus noticed them and suggested, incorrectly, that they were fossilized lentils. Really, they are fossil forams.

Forams, most of which are pinhead-sized, but some of which grow to inches across, are all, amazingly, single eukaryotic cells. This means again that, like you, they have a central nucleus containing chromosomes, protein-coated strands of DNA. You receive twenty-three from your mom, and twenty-three from your pop. They have mitochondria, also like you. They one up you, however: unlike you, many foram species have symbiotic inclusions, usually a single partnership with a photosynthetic being. This allows them not to work so much. Instead, they loll on the waves, harvesting in their specialty gardens. What this means is that they are chimeras, beings made of other beings. And in the case of the great art that is the Sphinx and the Pyramids, forty percent of the yellow limestone that makes up the Great Pyramids and Sphinx of Egypt consists of these nummulites, fossil forams that floated in the Tethys Sea during the Eocene, 56–34 million years ago.

Together with their amazing cousins, the radiating radiolaria and the diatomaceous earth-making diatoms, these foraminifera, an extremely widespread form of plankton, are jewels of the sea, using sunlight to take calcium, silicon, and other elements from the ocean and turning them into their own miniature microskeletons, their filigreed and fragile marine pillboxes, coins, and cathedrals. Like Picasso in his Blue Period delightfully constrained by limited materials to make superior artwork, life acts as an ingenious, but economically constrained artist, using what is available, assembling masterpieces from junk.

The elements incorporated into the psychedelic biosphere's bottom-up designs include silicon, the second most common element in Earth's crust after oxygen. Ninety percent of the minerals in Earth's crust are silicates. As silicon dioxide, this element composes glass, agate, tiger's eye, and rose quartz. The technology industry has found it useful, utilizing it in the silicon chips of our computers and cell phones. But if we see it in the stained glass of our churches, we should also appreciate it as the tiny, beautiful exteriors of diatoms (which can look like stained glass!), radiolaria, sponges, and other organisms. And silicon and calcium are only the tip of the biomineralizing iceberg. Life has been biomineralizing its environment for thousands of millions of years, turning its house into a home and its home into bodies as it assembles ever more complex bioarchitectures from its "nonliving" surroundings.

The notion that Earth is a rock with some life in it is part of our historical heritage and an example of Cartesian dualism: here, on this planetary *rock*, we have *matter*; while over here, in moving *organisms*, we have *life*. But life and its environment are so tightly wound that it is wrong to speak—or think—in this way, especially (I would say) if you are an artist. In fact, life and the environment form a single system, an energetic phenomenon of chemistry and movement, connected to the sun, at Earth's surface. "Life" is a kind of substance, one we feel from the inside, but which consists of cosmically available elements organized in regions of energy flow. Earth's surface is no more a rock with some life on it than you are a skeleton infested with cells. If you are sitting in a cafe or airport while reading this, look around at some of your fellow artworks and see if you like what life has done with its limited materials. I dare say that they are as impressive as the Watts Towers or any number of arresting entrées by Duchamp or others into conceptual, abstract, or performance art. Not to mention realism. From a materialistic standpoint, living matter is a peculiar moving mineral made mostly of water. It is, indeed, an impure form of water.

"I am as pure as the driven slush,"[8] said Tallulah Bankhead, and it is no insult to say that she recognized her sacred impurity. It is a complimelt (my neologism for a combined compliment and insult). The processes of liying organize many minerals on Earth's surface. Human teeth, for example, are converted toxic waste dumps: evolutionarily, my teeth derive from the need of marine cells to dump calcium wastc outside their cell membranes. Calcium is a mineral that will wreak havoc on normal cellular metabolism. Transporting this hazardous waste across cell lines in ancestral colonies of marine cells may well have been the basis for all present-day shell- and bone-making, including the apatite—a combination of calcium phosphate, calcium carbonate, calcium fluoride (not the same as the sodium fluoride added to drinking water)—and other compounds in a smile. Prevalent calcium minerals such as calcite, aragonite, carbonate, phosphate, halite, gypsum, and so forth are also the dominant media used in biomineralization. Opal, a semiprecious type of silica known for its iridescent play of colors, comes next. The magnetic mineral magnetite caps the teeth of chitons, but is also found inside the cells of bacteria, in swimming forms of algae, and in minute quantities in the brains of migratory fish, birds, sea turtles, and honeybees, where it may act as a compass. Opossum shrimp use needle-shaped fluorite crystals to avoid the light.

Like the found objects that a junk artist turns into beautiful works, calcium ions, once poisonous to marine cells, are now arranged in crystalline lattices to make shells and bones, including those of our ancestors with spines and central nervous systems—including, that is, those of the junk artist herself. Beautifully symmetrical marine microbes known as radiolarians deplete the oceans of amorphous silica and strontium to produce their ornate skeletons. The dried leaves of a New Zealand shrub consist of up to one percent nickel, a greater percentage than some mineral sources currently being mined. The concentration of vanadium in marine animals known as ascidians is one hundred times that of the surrounding sea, but why they collect it is a biological mystery.

8. Walter Winchell On Broadway, *Omaha World Herald*, September 3, 1941, quote page 6, column 5, www.quoteinvestigator.com/2013/09/20/driven-slush/#f+7256+1+1.

The nautilus, related to the squid and octopus, has a powerful aragonite "beak" capable of crushing bones. Its shell is also aragonite, while its balancing organ is formed of calcite crystals in a process of "pinpoint mineralization," and it has "kidney stones" made of phosphate minerals, inclusions found even in nautilus-type organisms whose kidneys have lost their function. These inclusions serve as reservoirs used to store the raw skeletal materials calcium and phosphorus within the organism. Mediating cell-to-cell interactions as part of the putative neurological basis for motor reflexes and thinking, calcium is lethal to cells in a free ionic state; but although calcium ions are 10,000 times more prevalent in the oceans than the poison cyanide, calcium itself has been incorporated into the very marrow of liýe, into its skeletons and shells, and into physiological processes ranging from blood-clotting to thinking.

Nor is all this work microscopic. Haptophytes, *Emiliana huxlei* (named after Thomas Huxley), form calcified round bodies known as coccolithophores. They are so prolific, copious, and profuse en masse that they form great marine clouds, for example off the coast of Brittany, visible from satellites as cottony masses in the ocean. Up close, the algal cells can be seen to sport skeletons that look like rounded, three-dimensional versions of Venetian blinds.

And this brings up an important point: liýe's beautiful and bizarre, sometimes absurd and horrific and ugly, and almost always interesting designs are never merely decorative. They always wed function to form in a way that makes use of a primordial excess. The coccolithophorids, for example, may be regulating their intake of light during photosynthesis with their Venetian blind-like exteriors. Exterior decoration, but highly functional. Nor is the functionality confined to their tiny bodies. It has been proposed that their gaseous production of dimethyl sulfide, a compound said to make the sea "smell like the sea," reacts in the atmosphere to form the nuclei of rain droplets. Thus, these tiny cells, and those like them, may play a major role in rain over the ocean and consequently help to regulate global weather and climate. Trees evapotranspire and produce terpenes. Cooling clouds are part of the Borneo jungle and Amazon rainforest. The physiology of their artistic bodies, in other words, connects to the physiology-like homeorrhetic stabilization of the global mean temperature (kept over geological time at a Goldilocks-like room temperature), the molecular composition of the atmosphere, the salinity of the oceans, and so forth. Because life consists not of closed but open systems, necessarily mediating flows of energy and matter—there is no life in a box—the organic artistry of cells and organisms made of cells bleeds onto the canvas at large.

Which leads to two subsidiary points, one involuting, like a Fibonacci-spaced slice of a foram shell, back to how I began. The first point and the second point are connected. The first point concerns my dear mother, Lynn Margulis. Her ability to focus scientific attention on the once-heretical notion that eukaryotic cell parts such as mitochondria and plastids were once free-living bacteria has its own prehistory. It is now proven, taught in textbooks, because those mitochondria and chloroplasts that have their own DNA were found to have that DNA sequenced in “smoking gun” arrangements far too close to free-living bacteria to be considered coincidence. They did it: they got together—being eaten by, parasexually infecting, and permanently mating with one another and archaea in a permanent corporal orgy far predating the sixties. That orgy is still going on, in, and as your body.

(Parenthetically let me remark that the rhetorical facility with which these ur-events spark reasonable, even salient metaphors—here the permanent orgy—is itself, to a writer like me fascinated by the process and act of writing, remarkable. I have sometimes envied visual artists, for example painters, for the freedom they have to be nonrepresentational. There is perhaps an argument for the existence of the writerly equivalent to impressionism, and to realism and minimalism, of course—and magic realism is a kind of surrealism—but abstract writing, cubist writing, expressionist writing, pointillism—there is a fluidity and diversity of styles in painting that writing has yet to tap.)

In 1956, when she was sixteen, she went into the field in Tepoztlán, Mexico, with Dr. Oscar Lewis. Later she did LSD when it was legal and also presented to generations of students the speech and life graphic drawings of royalist botanist Richard E. Schultes, cataloguer of more psychedelic plants in the Americas than any other person. And in her class, she revived Schultes’s ghost via digital interactive learning videos, and their analog forerunner Electrowriter, a quaint technology devised by Polaroid that recorded the lecturer's note-making hand as well as his or her voice, providing a unique, intimate experience of the lecturer's lesson.

It seems likely that teenage Lynn (she was eighteen at the time, Lynn Alexander, not yet Sagan, let alone Margulis) was awakened to a distinct view of nature, indigenous and flowing, in Mexico. In her house, she hung what one of her PhD geosciences students, Robin Kolnicki (an expert in chromosomal evolution in lemurs and bats, currently a biology professor at Framingham State University), described as "the dreams of a shaman." This artwork, with its colorful, woven, shape-shifting geometrical forms, may have been procured when she was conducting fieldwork as a student registered with the University of Illinois. According to her CV, in the Mexican village she studied "the modern doctor and the curandero" with Dr. Lewis. Another student of hers, Joanna Bybee, now a biotechnologist, wrote, "the swirling multi-colored forms in the artworks reminded me of the symbiotic bacteria that were her lifelong passion."

There may be something to this. Posting on social media, yet another student, Andre Khalil, pointed out that artist Shoshanah Dubiner's painting, *Endosymbiosis*, a psychedelic homage to Lynn's work on early liýe, strikingly recalled indigenous paintings inspired by ayahuasca—yage—the Amazonian dimethyltryptamine-delivering vine *Banisteris*, in combination with an MAO (monoamine oxidase)-inhibitor. These curandero paintings and Dubiner's homage highlight amazing swirling forms pulling and swarming in, through, and as discrete liýe-forms. During an interview with Sicilian-American science writer Dick Teresi that was published in a popular science magazine, Margulis said, "Do you want to believe that your sperm tails come from some spirochetes? Most men, most evolutionary biologists, don't. When they understand what I'm saying, they don't like it."[9] In this case she was talking about

9. Dick Teresi, "Discover Interview: Lynn Margulis Says She's Not Controversial, She's Right," *Discover*, June 17, 2011, www.discovermagazine.com/the-sciences/discover-interview-lynn-margulis-says-shes-not-controversial-shes-right.

an aspect of her theory of serial endosymbiosis that has not been accepted: the idea that spirochetes, known to feed upon, enter, and sometimes become permanently symbiotically attached to other cells (sometimes giving them powers of movement they would otherwise not possess), were the oldest (and thus most difficult to entangle) members of the symbiotic communities that evolved into eukaryotic cells.

To understand Margulis's work on the co-constitution of our bodies from different forms of bacteria and archaea, it is not enough to simply know about the origins of the plant plastids that paint nature green with cyanobacterial chlorophyll, or that the mitochondria you inherit from your mother's egg and that multiply in your muscles when you work out were once free-living respiring bacteria. You have also to know about her belief that there was an even more virulent and productive symbiosis in what would become the permanent orgy and artwork of eukaryotes. These were the most swirling of all swirly forms, the spirochetes. Spirochetes are both aerobic and anaerobic, that is, both poisoned by oxygen and not. There are structures in the balance organ in your inner ear, in the cilia of your oviduct, in your sperm tails, and in the sperm tails of ginkgo trees too, for that matter, that have not a genetic but a morphological structure—a kind of telephone dial of nine pairs of doubled tubes surrounding a central pair—that is very distinct. It strongly suggests that these structures, found in cilia when they are cut in cross section, share a common ancestry. Lynn speculated—call it a passionate hunch—that these structures were made possible, billions of years ago, by the matchmaking antics of spirochetes. For they are the fastest organisms in the microcosm, their corkscrew wriggling allowing them to race through viscous media from muds to saliva. One species, *Treponema pallidum,* causes syphilis and afflicted Franz Schubert, Arthur Schopenhauer, and Édouard Manet, among others. In *Pox: Genius, Madness, and the Mysteries of Syphilis,*[10] Deborah Hayden argues that syphilis is part of the story of Beethoven, Baudelaire, Maupassant, Flaubert, Oscar Wilde, James Joyce, and Nietzsche. Artists like these may have

10. Deborah Hayden, Pox: *Genius, Madness, and the Mysteries of Syphilis* (New York: Basic Books, 2003).

possessed a genius that was partly precipitated by spirochetes; the artists were, in a sense, more than human.

But for Lynn, the pathological spirochetoses of syphilis were the rare exception, not the rule. For her, eukaryosis or eukaryogenesis—the evolutionary production of eukaryotic cells—involved spirochetes in a process situated beyond sickness and health. Mitosis and meiosis, the dance of the chromosomes necessary for sexual reproduction as well as the microtubule-based electrochemical communications of neurons in the brain, were accomplished with liẏing remnants of vagrant spirochetes whose ancestors thrived billions of years ago in the Archaean muds, just as they do today—anaerobes hitchhiking through time, as it were, out of harm's way by maintaining a presence in animal tissues, safe from destruction by reactive O_2, oxygen gas. In an involutionary process more convoluted than anything yet in this essay, she believed that you can only understand this idea of the non-pathological role of spirochetes in evolution via the spirochetal remnants at the neurological basis of your own brain.

Like our industrial culture, which has commandeered energy at a global scale to run transportation devices from cars and trains to planes, but now uses electronic communication to transmit messages faster, the spirochetes settled down. Losing themselves, or parts of themselves, in the liẏing community, the symbiotic networks of combined eukaryotic cells we call the human body and brain, they now send messages rather than themselves.

To begin, I quoted Thoreau's statement about changing the quality of the day as the "highest of the arts" and Valerie Brown's musings on eukaryotes as a prokaryotic artform. We then discussed forams and other understudied, sometimes beautiful liẏeforms. The nucleated cells that evolved from bacteria and archaea became the first members of the protoctist kingdom. These beings that evolved into fungi, animals, and plants demonstrate wild variations on themes essential to human being. For example, animal-style multicellularity evolved in them. So did our kind of sexual reproduction, known as meiotic sex, which, despite anthropocentric appearances, has never been the norm, neither today nor in the course of evolutionary history. Some protists do not have mitochondria, even though they are eukaryotic. Some form slime on logs that dries to become brittle and orange before they reproduce. But all these eukaryotes, artistic on their own, are themselves, as Brown suggests above, a kind of artwork or creation of "lower forms," of the maligned germs without which we would not be here. Beauty made by scum. This is pretty fantastic but, spirochetes notwithstanding, it is also true—like the cryptic chimeric structure of the Sphinx.

Above, I wrote, "I would like to 'carve and paint the very atmosphere and medium through which we look,'" and added that, "In fact, you may be surprised to learn that some entities that you likely consider your deeply invidious inferiors have already done so." Now I am ready to make good on that Thoreauesque claim: Once upon a time, billions of years ago, Earth's sky was not yet blue. This is because liýe liberated such oxygen atoms from the hydrogen molecules to which they were molecularly bonded. The size of oxygen atoms is such as to scatter light of a blue wavelength. The liýe-forms responsible for this release of oxygen atoms were cyanobacteria. Their forerunners were purple sulfur bacteria that had thrived on hydrogen sulfide. The new water-using cyanobacteria took their hydrogen directly from the liquid medium in which they liýed. Breaking water's molecular bonds was difficult, but it saved early liýe from perishing and led to a liýe-form—a very successful one at that—that is arguably the dominant form of liýe on Earth today. And they were nicely named, considering what they did: cyanobacteria, with cyan denoting any of a range of colors in the blue/green range of the spectrum. These blue-green sun-lovers became trapped in the ancestors of plant cells, inaugurating the eternal salad days that land plants still enjoy. Language itself contains fossils, and thus it is that these beings are still sometimes referred to by their old name—"blue-green algae"—which tends to obscure what they are: bacteria, symbiotic aquatic oxygenators of a planet that, once upon a time, was a whiter shade of pale. It is an amazing story, all the more powerful for being both mythic and true:

Liýe gave Earth the blues.

Part Two: A Musical Phenomenology

Well, I have not really talked about animals, Heidegger, or Uexküll, but I think I have laid the groundwork. We are animals, of the chordate or Craniote phylum—it is not decided—whose members have notochords, sometimes spines, brains. There are another thirty-six odd animal phyla, but all the furry and zoo-ey charismatic megafauna are here: all the reptiles, fish, birds, mammals, the zebras, tigers, housecats, wildebeests, boors, and so on.

But, as we have seen, each such member of the Craniota phylum is not a flute solo, but a musical composition. Take yourself: if all "your" cells—that is the animal cells of your eukaryotic tissues, your blood, brain, lungs, skin, pancreas, stomach, and other organs—were instantly to disappear, say by a magical snap of my fingers, a ghost would still remain, as the late Hawaiian biologist Clair Folsome explains:

> What would remain would be a ghostly image, the skin outlined by a shimmer of bacteria, fungi, round worms, pinworms and various other microbial inhabitants. The gut would appear as a densely packed tube of anaerobic and aerobic bacteria, yeasts, and other microorganisms. Could one look in more detail, viruses of hundreds of kinds would be apparent throughout all tissues. We are far from unique. Any animal or plant would prove to be a similar seething zoo of microbes.[11]

And part of the interest here lies in a phenomenological dimension: these extras, these hangers-on to the host that is you, cannot be dismissed as completely other. They not only help you to metabolize your meat, synthesize your vitamins, and protect you from others of their kind—they also influence your mood, not just for the worse either, but sometimes for the better.

11. Clair Folsome, "Microbes," in *The Biosphere Catalogue*, ed. Tango P. Snyder (Fort Worth: Synergetic Press, 1985), 51–56; see also Aaron Bradshaw "Microbial Life," *Genealogy of the Posthuman*, accessed June 27, 2024, https://criticalposthumanism.net/microbial-life/.

Indeed, some may find that influencing your mood is good for *their* health, longevity, and reproduction.

To get a feel for this, take our chimeric friend, the cat, *Felis silvestris catus*. She is a peculiar being, big-eyed, slow then fast, sleek, a haughty hedonist with what seems like absolute independence, murderous indifference that commands our respect in the midst of domestic dependency. And it is here, in our houses, that her microbes become ours, and where what is part human becomes part feline, and vice versa. And the vector, one of an untold number linking us to other species, has only recently been characterized.

It is Toxo.

Toxoplasma gondii is a protist notorious for infecting pregnant mothers who may contract it from kitty litter. From the mother, *Toxoplasma* moves to the fetus, often devouring it and leading to a miscarriage. *Toxoplasma gondi* sexually reproduces in bodies of members of the Felidae family, notably house cats. But mice, usually afraid of cats, lose their fear when their brains become infected with *Toxoplasma*. Also, and curiously, they become sexually attracted to cat urine.

Toxoplasma infects large numbers of humans. You do not need to know Toxo is there to be affected by it. And *Toxoplasma* affects men and women differently. Infection in men correlates with enhanced risk-taking and jealousy. An index of such risk-taking behavior is provided by the fact that males in car and motorcycle accidents are more likely to test *Toxoplasma*-positive. Psychological profiles show that Toxo-men are more likely to be unfriendly, unsociable, and withdrawn. Even if bereft of obvious symptoms, men who carry *Toxoplasma* are less likely, relative to controls, to be found attractive to women.

Women are another story. Women with *Toxoplasma* are wild. They are more likely to be judged as outgoing, friendly, and conscientious—and promiscuous, at least when they are young. There is of course the complication of the stereotypical cat lady who lives alone (so to speak), surrounded by cats. The caricature of such a woman is not of someone outgoing. Perhaps Toxo's effects can dramatically alter with menopause.

However confounding, Toxo's effects seem real. *Toxoplasma* makes enzymes (tyrosine hydroxylase, phenylalanine hydroxylase) that alter brain levels of the neurotransmitter dopamine. Dopamine is involved in attention, sociability, and sleep. Cocaine and amphetamines work in large part by blocking the reuptake of dopamine in the brain. It has been theorized that dysfunctional dopamine regulation is linked to schizophrenia, and several antipsychotic drugs target dopamine receptors. Up to one-third of the world population is thought to be infected with Toxo, with an estimated infection rate of almost ninety percent in France—a statistic provisionally traced to the love among the French of rare beef, steak tartare, and beef *saignant*—"bleeding."

Czech scientist Jaroslav Flegr found via MRI scans that twelve of forty-four schizophrenia patients showed significant shrinkage of the cerebral cortex, but that the reduction in gray matter among the schizophrenics almost completely correlated with those who tested positive for *T. gondii*. Toxo, accounting for a range of effects and affects spanning from increased sexual attractiveness and feelings of well-being to full-on mental dysfunction, has phenomenological effects: it affects us from the inside, in our perception, as well as viewed from the outside, in our behavior. The Egyptian Sphinx is a sculpture, a monument. Its forams are real chimeras, but it is hard for us to know how they feel, or rather, felt. We, however, directly experience our own inner being. More and more this being, this inside, is showing itself to be chimerical, the result of not just us but also others. It is an ancient, ongoing, group production of liÿe-forms of many kinds hallucinating that they are one, impregnable and alone.

All these organisms and organelles have their effects on our multiplistic personality. *Candida albicans,* the yeast fungus which causes vaginal infections and perlèche, a cracking at the corners of the lips, may also change how we perceive and feel. Thriving on easily digestible sugars and carbs, it may lead hosts against their better judgment to crave beer, wine, cracker crumbs, and the other confections on which it feeds. One even wonders whether the wine-drinking revelers of Provence, the roving troubadours who sang and jested, might have sometimes used makeup as a way to cover their cracked lips. They are said to have invented romantic love in the thirteenth century. Performers, they may have resorted to such salves to treat lips that literally hurt when they smiled. Perhaps they drank wine and performed cunnilingus. Who knows? The show must go on. Candidiasis may thus be responsible for the origin of clowns and, more darkly, for the kitsch pictures of them hanging in many a lounge.

Human gut microbiota are not simply hangers-on but influence the timing of maturation for our intestinal cells, our internal nutrient supplies and distribution, our blood vessel growth, our immune systems, and the levels of cholesterol and other lipids in our blood. They also—partly because of the presence of neurons in the mammalian intestinal tract and the communication between gut and brain—influence human mood. Lab work with *Campylobacter jejuni* shows that this bacterium increases anxiety in mice, whereas the soil bacterium *Mycobacterium vacca* inside them cheers them up. In people, it has been suggested that yogurt with liýe cultures, for example, with bifidobacteria, improves our sense of well-being.

Now we come briefly to what might be called the biosemiotic inside, the Uexküllian music, of what is ultimately the single side of a Möbius strip of liýing being.

Martin Heidegger, a most influential philosopher of the twentieth century, developed his deep and unique discourse on the essence of the human being as being there, *Dasein,* based in part on his teacher, Edmund Husserl, to whom he dedicated *Being and Time.*[12] Husserl, a mathematician, established phenomenology, the attempt to look scientifically at the reality of inner phenomena, including the pure forms of the mind. Husserl's transcendental reduction tried to remove what is extraneous to get at the essence of existence, and morphed, in a sense, into Heidegger's existential analytic.

But Heidegger—who did not support Husserl, who was Jewish, under the Nazi regime—was also deeply influenced by Jakob von Uexküll. Not only does Heidegger reference the ethologist more than any other single scientist, but one can argue that *Dasein* is an

12. Martin Heidegger, *Being and Time* (New York: SUNY Press, 2010).

unnecessary, anthropocentric metastasis of von Uexküll's more basic idea.

That idea was simple. It was that every organism, to varying degrees and with various stimuli, senses, and desires, has its own inner world. The Estonian-born Uexküll called an animal's perceptual lifeworld *Umwelt*. Significant things trigger chains of events, sometimes spelling the difference between liyfe and death. Organisms are aware. They are looking for things. They have search images. Cyclical beings, they are not just reactive machines but projective beings.

Uexküll, a biologist and neo-Kantian, shamanizes himself into the lifeworlds of other, nonhuman beings. He tells us what it is like to be a dog, unbothered by the incline of a grassy plain that would impede us, with only two legs. He tells us of animals who move their legs and those whose legs move them. Incredibly, in *A Foray into the Worlds of Animals and Humans*,[13] he shows us what a scallop would see looking onto a Bavarian street. I hope I have that right. Uexküll shows us the sea urchin extending its spines to the stimulus of passing ships and cloud, which the sea creature misinterprets as a potentially deadly predator fish. He intuits the plight of the fly, its vision unable to resolve the strangling strands of the spider's web, the jackdaw fooled by a cat carrying a rag. Even the world of the blind, deaf tick, sensing mammals by the slight amount of butyric acid their bodies give off, is uncovered by Uexküll's shamanic *Umwelt*-vaulting.[14] Deleuze and Guattari suggest that some people's *Umwelten* are not so advanced.[15]

But in addition to Uexküll's stick-searching dogs, hypothesis-generating scientists, and starfish-avoiding scallops, there are an estimated 10 to 30 million extant species: water scorpions with built-in fathometers sensing hydrostatic pressure gradients, plants with gravity sensors, algae-perceiving barium sulfate and calcium

13. Jakob von Uexküll, *A Foray into the Worlds of Animals and Humans: With a Theory of Meaning*, trans. Joseph D. O'Neil (Minneapolis: University of Minnesota Press, 2010).

14. Ibid.

15. Gilles Deleuze and Félix Guattari, *A Thousand Plateaus: Capitalism and Schizophrenia*, trans. Brian Massumi (Minneapolis: University of Minnesota Press, 1987), 257.

ions, fish that gauge the amplitude and frequency of turbulent waters with dipole electrostatic field generator-and-sensors, magneto-sensitive bacteria, homing pigeons, polarized light-detecting bees whose peregrinations are not impeded by clouds, and male silkworm moths sensing female silkworm moths miles away.

In a move arguably similar to the Nazi rhetoric of purity, Heidegger cleansed this diverse Uexküllian multispecies music. But Heidegger turned Uexküll's notion of an organism's lifeworld, its *Umwelt,* into something solely human. Heidegger argues that man exclusively dwells in the house of language, that Hölderlin was wrong in thinking that animals, even a lark flying across the sky, have access to "the open"—that is, Being as a whole rather than an environment. Heidegger asserts that the human hand is vastly different from a monkey's paw. Heidegger considers all nonhuman animals together. Apparently almost any animal will do—he chooses a bee—to stand in for the essence of "the animal." The bee, Heidegger asserts, referencing an experiment, will continue to drink even if the connection from the mouth and the stomach is severed. It seems that for post-Catholic Heidegger, the bee is a Cartesian automaton. Or something close to it. Heidegger says that man has world, a rock has no world, and animals are poor in world. Animals are separated from us language-dwellers by an "abyss." So, Heidegger takes Uexküll in a distinct direction, running with his important exploration of meaning-making, purposeful behavior, and the inner worlds of various organisms, but then applying them almost solely to man.

Interestingly, both biosemioticians, who study biosemiosis, the "sign action in living systems"—which in theory applies to all organisms—also trace back to Uexküll. Biosemiotics founder Thomas Sebeok called Uexküll a "proto-biosemiotician." For most biosemioticians, language systems and life completely overlap, as attested to by talk of "messages," "reading," "translation," and "code" in modern genetics, which, however, does not explicitly have a theory of the sign. Many biosemioticians are big on the grandfather of American pragmatism, Charles S. Peirce, who argued that the sign is triadic, meaning that you cannot just have a message, you have to have someone or thing that sends it and a recipient that interprets it. But who or what is sending the message that is DNA? Who, or what is receiving it?

I would argue, as Heidegger's bee example tries to show, that part of the reluctance to grant language or language-like processes to other beings has to do with a crypto-Cartesianism. *We* are the agents, the ones with free will, the responders, who see the whole, the godly even in the absence of God; "the animal" just sees what is in front of his nose, the parts. Whether or not it is right, it certainly is a human exceptionalism.

Although there is no time to delve into it here, I think that this is a dubious proposition. We can "bracket," to use Husserlian phenomenological terminology, the question of our freedom—forams make aesthetic choices, but are they free?—and still appreciate the endless richness of our synesthetic senses. Invoking "transensual, timeless"[16] knowledge that allows organisms without human foresight to act in ways that match present action to future needs, Uexküll speaks of a musician-like "composer" of awareness, who is "aware" and can "shape future life-requirements" with a "master's hand."[17]

While this is still too anthropomorphic for me, I like the idea of our sensibilities, our inner senses, as instruments that are played rather than something that we play. Our awareness of this process is itself a kind of knowledge, a freedom. It is perhaps a fortuitous coincidence that the Greek word for organism, *organon,* also means instrument. Many pollinating insects detect flowering plants invisible to those who cannot see in the ultraviolet range below four hundred nanometers in wavelength. At the other end of the spectrum, pit vipers such as rattlesnakes detect infrared radiation (heat) too subtle for us to notice. Bats determine the size, location, density, and movement of prey such as fruit flies one hundred feet away in a pitch-black cave using sonar, emitting through their mouths and nostrils ultrasound vibrating at frequencies of some 100,000 cycles per second, about five times what we can hear. Dolphins echolocate in the water by making click sounds, and humpback whales sing to each other thousands of miles across

16. Von Uexküll, *A Foray into the Worlds of Animals and Humans.*

17. Ibid.

the ocean, making songs that completely change over a five-year period, using some of the same rules that human composers do.

We know what it is like to be at a rock concert. But what is it like to be a hawk soaring over a rock, a heron fishing with an iridescent beetle, a swarm of bees in a field of ultraviolet-patterned flowers? The Jesus lizard walking on water? The pebble toad, *Oreophrynella nigra,* curling up and bouncing down a Venezuelan mountain? The queen bee laying a thousand eggs in her fragrant maze? The female mosquito finally sucking blood from a host during a muggy dusk? Individual cells within that host suffering from a delusion of absolute individuality? The water lilies in the pond where dived the fat bullfrog whose croak filled the sounds of the rustling forest the night before? The forest itself, in its slow, chemically complex growth, its subterranean nuclei-trading networks of mycelial tendrils, its bacterially diverse metabolism, able to breathe arsenic and sulfur, to grow and merge? The biological world is a vast colossus of differentiable and merging sensibilities. … All of these and each of us, one might argue, to *whatever* extent we actually control our actions, are spectators of the great show on Earth. Each of us is, in a way, like an instrument in a billion-year-old symphony, making vibrations and emotions and Uexküllian music, for ourselves and one another, if not, as the post-Kantian biologist seems to have implied, as localized beings in the perception of a multidimensional, immanent being seeing itself, at least partially, through its manifold perceivers.

Coda

We carve the air not just through art but through perception and theory. In her essay "The Beauty of the World," Sharon E. Kingsland says that, for G. Evelyn Hutchinson, a philosopho-scientific polymath and one of the founders of modern ecology,

> contemplative values need not be nurtured. … But what did Hutchinson mean by "beauty"? He explained by relating an anecdote about an experience he had while walking down the drive of his house. On that occasion he spied a brilliant patch

> of red, which drew his attention and puzzled him: "In a second or two I realized that a pair of scarlet tanagers was mating on a piece of broken root conveniently left by a neighbor's somewhat inconsequential bulldozer; the female was sitting inconspicuously on the root, the male maintaining his position on her by a rapid fluttering of his black and hardly visible wings which tended to vibrate his entire body." He reflected that the sight was strikingly beautiful and that it gave him a sense of pleasure to realize this.[18]

The sight of the red patch that turned out to be a pair of mating tanagers spurred Hutchinson to think of "an amorous and beautiful seventeenth-century song"; it conjured forth "religious and psychoanalytic connections," and the "color itself reminded him of specimens of Central American tanagers that he had seen in a museum, which caused him to think about the evolution of these birds."[19]

When I think of Hutchinson's tanagers, I think of briefly meeting him at the Great Hall of Dinosaurs at Yale's Peabody Museum of Natural History in the early 1980s; I think of the color red and how the eye alights to it in art, where it should be used judiciously; I think of the ecstatic intensity of the sex act, a kind of ellipsis in the liyfe sentence of human identity; I think of Georges Bataille saying that the sexual act in space is what the tiger is in time.[20] I think of Heidegger returning in 1938 to the question of freedom, partly in response to Friedrich Schelling, who was himself responding to the dense, articulate, multifaceted dismissal, the post-Cartesian pre-postmodern deconstruction of free will by Spinoza. Spinoza did not believe in freedom as volition, but he did believe in freedom, strongly, as a political necessity as well as a kind of intellectual love of the cosmos, a widening of the contemplative spirit. For

18. Sharon E. Kingsland, "The Beauty of the World: Evelyn Hutchinson's Vision of Science," in *The Art of Ecology: Writings of G. Evelyn Hutchinson*, ed. David K. Skelly, David M. Post, and Melinda D. Smith (New Haven: Yale University Press, 2010), 4.

19. Ibid.

20. Georges Bataille, *L'Anus Solaire*, illustrations by André Masson (Paris: Éditions de la Galerie Simon, 1931).

me, as for Spinoza, agency may be an illusion. But that does not take anything away from art or experience. It is as if we are on a ride, instruments being played, even as we think we are in control.

Hutchinson was inspired by his friend, the novelist, Rebecca West. West wrote an essay about art, "The Strange Necessity."[21] As Sharon E. Kingsland writes in her introduction to *The Art of Ecology: The Writings of G. Evelyn Hutchinson,* Hutchinson's red tanagers "illustrate how the seemingly simple and direct experience" is "conceptually enriched by so many kinds of association that … it … is"—and here she quotes him—"'essentially an art form.'"[22] ∞

Digital Content 04 Lynn Margulis, Bruce Clarke, Dorion Sagan, and Spherical, *untitled*, 2022.

21. Rebecca West, "The Strange Necessity," in *The Strange Necessity: Essays by Rebecca West* (Garden City: Doubleday, Doran & Company, 1928), 1–213.

22. Kingsland, "The Beauty of the World," 4.

Walk through *Holobiont: Life is Other*

01 Thomas Feuerstein, Jens Hauser, Judith Reichart, and Lucie Strecker (curators)
Angewandte Interdisciplinary Lab, (AIL) Vienna 2022

With contributions by Art Orienté Objet; Irini Athanassakis; David Berry; Guy Ben-Ary, Julia Borovaya; Adam W. Brown; Juan M. Castro and Akihiro Kubota; Tagny Duff; Thomas Feuerstein; Karmen Franinović; Ana María Gómez López; Luis Hernan, Pei-Ying Lin, and Carolina Ramirez-Figueroa; Hideo Iwasaki; Henrik Plenge Jakobsen; Eduardo Kac; Roman Kirschner; Lynn Margulis, Dorion Sagan, Bruce Clarke, and Spherical; Yann Marussich; Agnes Meyer-Brandis; ORLAN; Špela Petrič; Chris Salter; Maja Smrekar; Klaus Spiess; Lucie Strecker; Tina Tarpgaard; Paul Vanouse; M R Vishnuprasad; Peter Weibel; KT Zakravsky; Adam Zaretsky and the authors of the special issue "On Micorperformativity," ***Performance Research: A Journal of the Performing Arts*** **25, no. 3 (2020).**

02 Adam W. Brown
The Great Work of the Metal Lover

2012, video and tableaux, ***Holobiont: Life is Other,*** **Angewandte Interdisciplinary Lab, Vienna, 2022.**

03 Thomas Feuerstein
GREEN HYDRA

2021, installation, ***Holobiont: Life is Other,*** **Angewandte Interdisciplinary Lab, Vienna, 2022.**

04 Lynn Margulis, Bruce Clarke, Dorion Sagan, and Spherical
untitled

2021, tableaux, ***Holobiont: Life is Other,*** **Angewandte Interdisciplinary Lab, Vienna, 2022.**

05 Paul Vanouse
Labor

2019, documentary video, *Holobiont: Life is Other*, Angewandte Interdisciplinary Lab, Vienna, 2022.

06 Klaus Spiess, Ulla Rauter, Emanuel Gollob, and Lucie Strecker
ECOLAIA

2022, video and installation, *Holobiont: Life is Other*, Angewandte Interdisciplinary Lab, Vienna, 2022.

07 Maja Smrekar
K-9_topology: HYBRID FAMILY

2016, video and tabelaux, *Holobiont: Life is Other*, Angewandte Interdisciplinary Lab, Vienna, 2022.

08 Maja Smrekar
K-9_topology: Through the Lens of a Long-Term Care of an Artwork

2022, video essay, *Holobiont: Life is Other*, Angewandte Interdisciplinary Lab, Vienna, 2022.

09 Sylvia Eckermann and Gerald Nestler
Future of Demonstration

2023, documentary video, *Planetary Skins*, Gallery Elisabeth & Klaus Thoman, Innsbruck.

2024, documentary video, *Like a Ray in Search of its Mirror*, Forum Frohner, Krems.

Acknowledgements

This book, *Life Is Other: A/Biotic Entanglements in Art and Curating*, would not have been possible without the invaluable contributions of our esteemed colleagues and their resolutely transdisciplinary minds. Our collaborative efforts both on the evolving exhibition formats and this publication have been marked by intensive teamwork, mutual support, constructive criticism, and a shared commitment to radical openness in exploring these complex themes at the threshold of art, science, and post-anthropocentric cultural theory.

All of the contributors who participated in this project stand out due to their dedication to fostering creative, bold, and critical discourse on the pressing issues of coexistence in the times of ecological crisis that we are collectively facing. We extend our heartfelt thanks to all those who have supported this book, putting aside their own economic interests for the sake of the project and allowing us to bring our artistic research to fruition. We hope that the volume will serve as a catalyst for further entanglements and the emergence of future interconnected exhibitions.

We would like to express our sincere gratitude to the authors and artists involved in the book for their invaluable contributions and for putting their trust in us as editors. The teams from both exhibition venues, Magazin 4, Bregenz, and Angewandte Interdisciplinary Lab (AIL), Vienna, realized our show in the most inspiring way. Our special thanks go to Judith Reichart, Head of Cultural Affairs in Bregenz, and her dedicated assistant, Julia Hagspiel. For reworking the exhibition for the Vienna venue and hosting a number of associated events, we are deeply grateful to Alexandra Graupner, Elisabeth Falkensteiner, Eva Weber, and Nico Wind. Wolfgang Fiel's exhibition design made a significant contribution to the success of the show. We would additionally like to thank Jürgen Weishäupl and Lukas Böckle for supplying us with pioneering new technologies for a digital archive. We would also like to acknowledge the scientific support provided by David Berry and Silvia Bulgheresi, whose talent for communicating complex biological concepts to a broader audience without over-simplifying them encouraged us to embark upon this challenging publication. The proofreading and copyediting were undertaken by Lydia J. White with the utmost care and precision. We are highly grateful for her dedication and patience. Simona Koch, who came up with the book design, was at the heart of this endeavor, infusing joy, inspiration, and her conceptual gaze throughout the creation process. We would also like to thank Barbara Wimmer, whose reliable production management experience was indispensable to this project.

This exhibition and the book project would not have been possible without the support of the Austrian Science Fund (FWF) and Division IV—Arts and Culture of the Republic of Austria's Federal Ministry for Arts, Culture, the Civil Service, and Sports (BMKOES). We are honored to have had the support of our sponsors ML Manna Laaz BeteiligungsgmbH, Vienna, and Maggie Luan for this publication.

Finally, we would like to extend our thanks to the Rector of the University of Applied Arts Vienna, Petra Schaper Rinkel, for allowing the publication to become part of the *Edition Angewandte* series. It is a visible testament to the university's rich practice of combining in-house and external expertise, fostering a continuous, open discussion.

as Feuerstein, Jens Hauser, and
Strecker (Eds.)

ublication is a result of the exhibi-
Iolobiont, curated by Thomas
stein, Jens Hauser, Judith Reichart,
ucie Strecker, and held at Magazin 4,
genz (April 17–July 7, 2021). The
ition was adapted and expanded for
ngewandte Interdisciplinary Lab
, A-Vienna (October 5, 2022–January
23)with support from Elise Richter
Project V501, funded by the Austrian
ce Fund (FWF).

ed with the support of Divison
rts and Cultures of the Republic
stria's Federal Ministry for Arts,
res, the Civil Service, and Sports
OES) as well as ML Manna Laaz
ligungsgmbH and Maggie Luan,
nna

ct Management "Edition
wandte" on behalf of the University of
ied Arts Vienna:
ara Wimmer, A-Vienna

editing/proofreading:
J. White, NZ-Wellington

design, layout, cover design,
ypography:
na Koch, A-Vienna

e editing:
an Regl, A-Vienna

R 3D scans and metadata
agement:
s Böckle, NEST Agency, A-Vienna

ing:
hausen, die Buchmarke der Gerin
k GmbH, A-Wolkersdorf

r:
ken Lynx Rough 100 g/m²

face:
ce Serif, designed by Frank
hammer, from Adobe Originals

The text "Labor: The Post-Anthropocentric Body at Work," by Paul Vanouse, pp. 28-33, has been adapted and reprinted with the kind permission of Routledge.
© *Performance Research: A Journal of the Performing Arts,* 2020, 25 (3)

Library of Congress Control Number:
2024947640

Bibliographic information published by the German National Library
The German National Library lists this publication in the Deutsche Nationalbibliografie; detailed bibliographic data are available on the Internet at http://dnb.dnb.de.

ISSN 1866-248X
ISBN 978-3-68924-133-9
e-ISBN (PDF) 978-3-68924-011-0

Federal Ministry
Republic of Austria
Arts, Culture,
Civil Service and Sport

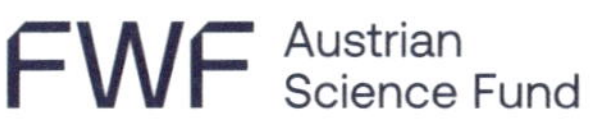

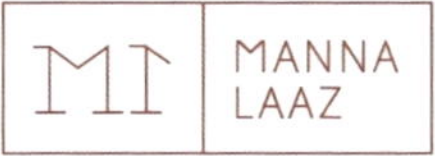

Imprint

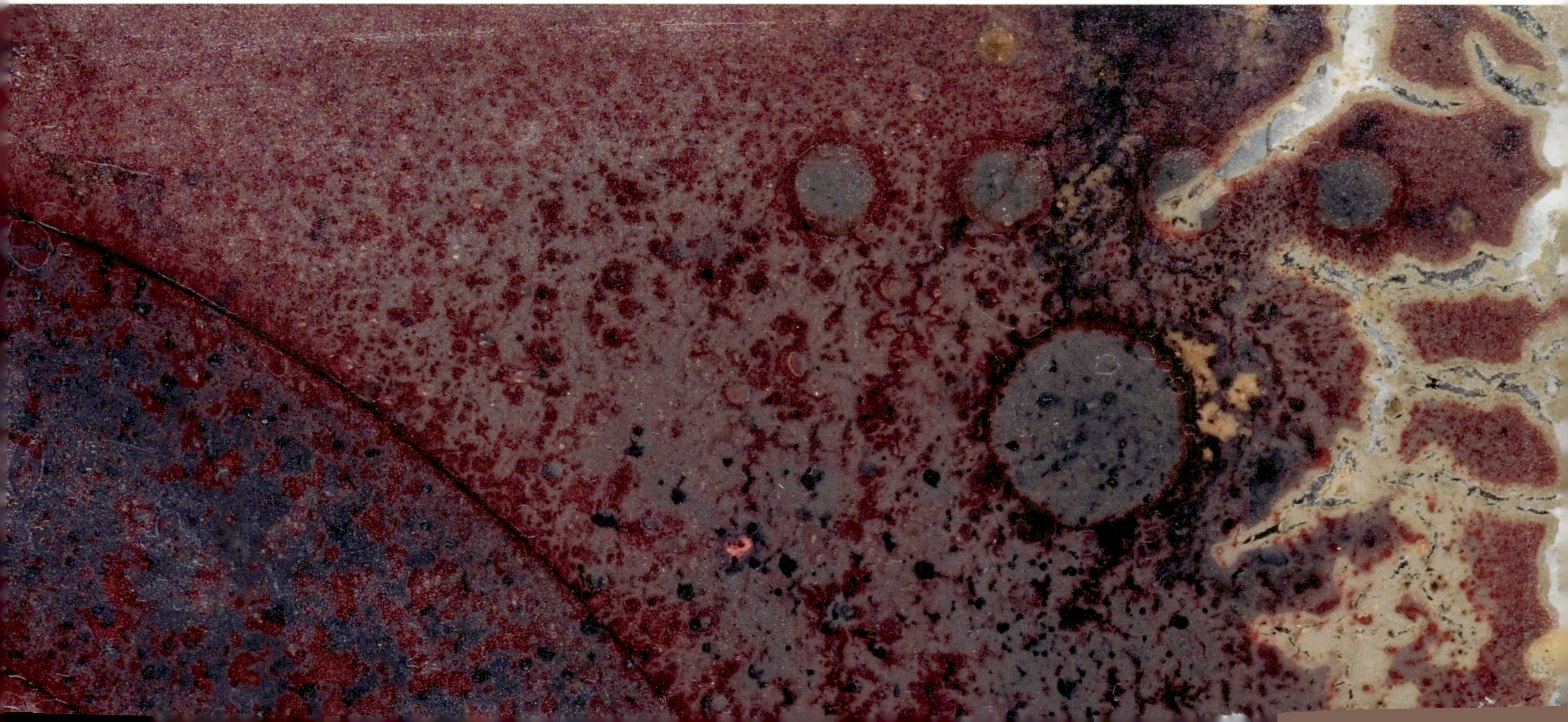

Fig. 83 Eduardo Kac, *Hullabaloo*, biotope, 2006. Collection Alfredo Herzog da Silva, São Paulo. © Eduardo Kac